Intersex in Christ

Intersex in Christ

AMBIGUOUS BIOLOGY
AND THE GOSPEL

Jennifer Anne Cox

Foreword by
Rev. Sandra Basham

CASCADE *Books* · Eugene, Oregon

INTERSEX IN CHRIST
Ambiguous Biology and the Gospel

Cascade Books
An Imprint of Wipf and Stock Publishers
199 W. 8th Ave., Suite 3
Eugene, OR 97401

www.wipfandstock.com

PAPERBACK ISBN: 978-1-5326-1845-1
HARDCOVER ISBN: 978-1-4982-4402-2
EBOOK ISBN: 978-1-4982-4401-5

Cataloguing-in-Publication data:

Names: Cox, Jennifer Anne. | foreword by Rev. Sandra Basham.

Title: Intersex in Christ : Ambiguous biology and the gospel / Jennifer Anne Cox ; foreword by Rev. Sandra Basham.

Description: Eugene, OR: Cascade Books, 2018 | Includes bibliographical references.

Identifiers: ISBN 978-1-5326-1845-1 (paperback) | ISBN 978-1-4982-4402-2 (hardcover) | ISBN 978-1-4982-4401-5 (ebook)

Subjects: LCSH: Sex differences—Religious aspects—Christianity.

Classification: BT708 .C68 2018 (print) | BT708 (ebook)

Manufactured in the U.S.A.   MAY 27, 2018

Contents

Foreword

A THEOLOGICAL BOOK ABOUT intersex people; how thought provoking—and necessary. How does the church in the twenty-first century relate to and minister to intersex people? People whose "image and likeness" of the creative and benevolent God may appear to some Christians to be marred?

What scriptural guidelines and truths are gleaned and debated on this topic in Christian academic or devotional literature? This book is utterly unique. There are sexology and medical books about intersex people, but not one that weaves the Scriptures through its warp and weft, deeply examining the "image and likeness" of God in intersex people and validating their worth and inclusion in God's kingdom and church.

This book is thought-provoking, compassionate, respectful, and challenging. It is a challenge the church needs to face. All people are afraid of what they do not understand. Christians are no exception. This book will help you to challenge your own thinking and theology about intersex people, both inside and outside of the church.

I have the privilege of working with the author, Dr. Jenny Cox, at Tabor College, Perth. I strongly encouraged her to write this book, for theologians, for the church, for pastors, for Christians, for sexologists like myself, for everybody. I sincerely applaud her effort.

Rev. Sandra Basham
Senior Lecturer in Humanities and Social Sciences
Tabor College of Higher Education (Perth, Western Australia).
Cert IV TAA, DipACEd, DipCC&FT, BEd, MForSx, PhD (candidate).

Acknowledgements

There are many people who I want to thank, because they have helped me to bring this book into being. Thank you to my husband, who financially supports our household, leaving me time to write books. Thanks to Dr. Stuart Devenish of Tabor (Adelaide), who has encouraged me and given both writing and publishing advice. Thanks to Sandra Basham, my colleague in Tabor (Perth), who has been my cheer squad as I worked through this topic. Thanks to Tabor College of Higher Education and to the College's Centre for the Promotion of Church Health, which provided a financial grant to help me complete the book. And lastly, thanks to all the people in my life who have given me prayer support and encouragement throughout.

Abbreviations

5-ARD	5-alpha reductase deficiency
AIS	androgen insensitivity syndrome
AISSGA	Androgen Insensitivity Syndrome Support Group Australia
AMH	anti-Müllerian hormone
CAH	congenital adrenal hyperplasia
CAIS	complete androgen insensitivity syndrome
DHT	dihydrotestosterone
DSD	disorders of sex development
DSD	disorders of sex differentiation
ISNA	Intersex Society of North America
LGBTQ	lesbian, gay, bisexual, transgender, queer
OIIA	Organization Intersex International Australia
PAIS	partial androgen insensitivity syndrome
UKIA	United Kingdom Intersex Association

Introduction

Beginnings

Out in the Open

INTERSEX IS "COMING OUT of the closet." More and more intersex people are telling their stories. One prominent intersex Australian is Tony Briffa, the former mayor of Hobson Bay. Tony grew up as Antoinette and later found out that he/she is intersex. Tony is now quite open about being "not male or female, but both."[1] But Tony is not alone. In 2014 the ABC (Australian Broadcasting Corporation) interviewed two intersex persons—Shon Klose and Georgie Yovanovic. At age sixteen Shon discovered that she has Mayer-Rokitansky-Kter-Hauser syndrome, meaning she lacks any internal reproductive organs. Shon was pressured to have surgery to create a vagina but never offered counseling.[2] Georgie also has a sad story to tell. Born with male genitalia but in every other way looking like a girl, Georgie was abused by her father, who refused to let Georgie 4ct in any way like a girl. But at puberty Georgie grew breasts and was forced into surgery. Georgie has unusual sex chromosomes, XXY instead of the normal XX or XY.[3] All of these people would have been invisible not many years ago.

Intersex organizations now exist in many countries to give intersex people both a voice and support. Australia has two such organizations. Organization Intersex International Australia (OIIA) provides support to

1. Briffa, "About Tony."
2. Sleath, "I Am Intersex: Shon Klose's Story."
3. Sleath, "I Am Intersex: Georgie Yovanovic's Story."

1

intersex persons and their families, educates others, and contributes to government policy on intersex issues.[4] Androgen Insensitivity Syndrome Support Group Australia (AISSGA) helps people with intersex variations meet other intersex people, supports parents, enables openness about AIS, and promotes research.[5] Support groups like these are a relatively recent phenomenon. Possibly the most prominent support group—Intersex Society of North America (ISNA)—did not come into existence until 1993.[6] Before the existence of support groups like these, intersex was a well guarded secret to all except doctors and parents, often being hidden even from the intersex persons themselves.

Why Write an Evangelical Response to Intersex?

Although medical, psychological, and sociological literature about intersex exists and is growing, there has not been a great deal of theological engagement with intersex. A few articles[7] and book chapters[8] have been written. Book-length liberal Christian responses exist.[9] Yet an evangelical Christian response, which considers intersex through the lens of Christ, his person and work, is needed.

Some quotations will underscore the need for an informed evangelical response to intersex. Sally Gross, an intersex woman who was formerly a Catholic priest, explains the "logic" of why her baptism was not valid:

> The argument, which was put to me by conspicuously pious, intelligent, theologically sophisticated but fundamentalist Christians of my acquaintance, is roughly as follows. Gen[esis] 1:27 states that from the beginning of creation, God made each given member of the human species either male or female, and not both or neither. Thus, determinate maleness or determinate femaleness is the mark, above all else, of what it is to be created human. Validity

4. "About OII Australia."

5. AISSGA, "Peer Support, Information and Advocacy for Intersex People and Their Families."

6. ISNA existed from 1993 to 2008 (ISNA, "Our Mission").

7. Cornwall, "British Intersex Christians' Accounts"; Cornwall, "Telling Stories about Intersex and Christianity"; Gross, "Intersexuality and Scripture"; Jung, "Christianity and Human Sexual Polymorphism"; Lebacqz, "Difference or Defect?"

8. Burk, *What Is the Meaning of Sex?*, 19–20, 160–83; Weerakoon, *Teen Sex by the Book*, Chapter 9.

9. Cornwall, *Sex and Uncertainty*; Mollenkott, *Omnigender*.

of baptism is reserved for those who are human: one could immerse or sprinkle a dog, cat or tin of tuna, sincerely intending to baptize these, while uttering the formula of baptism, but no attempt to baptize these could ever be valid because dogs, cats and tins of tuna are not the kinds of thing which can be baptized and only human beings can be baptized. Since I am intersexed and my congenital physical sex has been found to be as ambiguous as it could be, I do not satisfy the divine criterion for humanness, which requires that one objectively be either determinately male or determinately female. It follows that, like dogs, cats and tins of tuna, I am not the kind of thing which could have been baptized validly.[10]

Gross's statement demonstrates how ignorance can cause so much pain to people for whom Christ died.

An Australian intersex person bewails the failure of religious people to exercise concern for intersex individuals:

In the one hundred and fifty years since Herculine Barbin was born not one single religious organisation has stood up for the rights of intersex people.

Not one religious organisation has argued for our equal rights and social inclusion.

Not one religious organisation provides intersex-specific pastoral care.

Not one religious organisation provides resources or support for intersex people.[11]

Although not necessarily directed at Christians alone, this statement challenges the church to act and to speak up for intersex people.

A third statement highlights the need for theological work by evangelicals. "I'm intersex. Evangelicals don't have a category for me, so there's no real place for me in their church."[12] Together these call for a response of compassion and pastoral concern for the needs of intersex individuals. Such a pastoral response needs to be informed by a carefully articulated theology.

10. Gross, "Intersexuality and Scripture," 70 fn 7.
11. Wilson, "Intersex and Religion," 44. Formatting original.
12. Adrian quoted in Evans, "The False Gospel of Gender Binaries."

An Evangelical Approach

There is one substantial evangelical response to intersex. Megan Defranza's *Sex Difference in Christian Theology* is a positive and helpful exploration of intersex. Educated in both evangelical and Roman Catholic theology, hers is a conservative theological stance. She opens up a positive space for intersex people within the church. If an evangelical theology of intersex already exists, then why is another book needed? There are many more issues to be addressed than was possible in DeFranza's book, including medical treatment of intersex, identity formation, shame, sexual abuse, inclusion in church, the purpose of gender, and sexual ethics for intersex persons. I offer some perspectives on intersex issues that DeFranza did not.

I began researching this book because a friend asked me what I thought about intersex people. At that time I could offer no opinion, theological or otherwise, because I was utterly ignorant about intersex. Once I began researching, I realized that the issues that surround intersex require more than a casual response. I must admit that I am neither an expert in sexuality or gender studies, nor a biologist or psychologist or sociologist, and I have no medical training. However, I have drawn from studies in sexuality and gender, biology, psychology, and sociology, along with medical discussions. I have no personal experience of intersex. I am unambiguously sexed and heterosexual, married to the same man for more than thirty years. But I have sought to find out as much as I could about what intersex people have to say about having an intersex variation.

The issues I address in this volume were determined by what intersex people have stated as significant. I discovered these by reading interviews with intersex persons and studies done on intersex individuals, through blogs and the websites of intersex societies. Although I have used medical, sociological, and psychological literature to try to understand what intersex is and what issues surround it, this book is primarily a work of theology that draws on biblical themes. Being a practicing Christian of more than thirty-five years and an evangelical theologian, I know that human significance is not found in medicine, sociology, or psychology, but rather in the person of Jesus Christ. It is in him that we find our worth, dignity, and purpose.

My approach to a theology of intersex does not involve trying to find intersex individuals in the Bible. Nor am I concerned to look for passages that might be a basis for arguments against genital surgery. This kind of approach is too narrow and can only bear minimal fruit. Instead I have taken a broader approach that puts Jesus Christ as the theological center.

The book explores significant aspects of God's work in the world, generally following the order of events in the biblical story. It begins with creation and fall, and the significance of human sexuality. Then it considers the consequences of the life and death of Jesus. The story of salvation is completed in the resurrection of the dead. Within this broader understanding of the gospel story as the framework, it is possible to see that although the Bible does not directly mention intersex, it most certainly has much to say about the topic.

An evangelical approach will be both critically realistic and optimistically focused on God's ultimate goal for humanity. We must love people as they are, while also considering what they will become. God does not love us only when we have become perfect examples of humanity, but rather he loves us because we are his creatures and *so that* we can become what he intends us to be. An evangelical response to intersex, then, must do several things. It must first acknowledge what intersex actually is and the issues that surround it. Second, it must work with the tension that world is good because it is created by God but it is now distorted through sin. Third, an evangelical theology must center on Jesus Christ, true God and true human, source of salvation, and the one to whom we must all become conformed. Because of Christ, God's grace and mercy permeates human existence, even in the midst of decay and evil. Lastly, an evangelical response will not ignore the final state of humanity. However un-ideal our humanity is in the present, it will become something profoundly glorious when Jesus Christ returns. It is within these parameters that I have set my theological sights.

Some Definitions

Since this book addresses issues do to with sex and gender it is important to provide some basic definitions of these before proceeding. "Sex" generally refers to biological sex, that is, the bodily expression of sex. Biological sex is defined by chromosomes, genes, hormones, internal genitalia (gonads), and external genitalia. "Gender," on the other hand, refers to thoughts and actions rather than biology. "Gender identity" entails what a person thinks about himself or herself. "Gender expression" includes gender roles, behaviors, dress, and interaction with others. "Sexual orientation" refers to whether a person is heterosexual, homosexual, bisexual, or asexual. This is about attraction to other persons. Biological sex, gender, and sexual orientation are often related but should not be confused or fused together.

Although I fully acknowledge the different aspects of sexuality described above, throughout the book I have used the word "sexuality" in a very broad way. Although sexuality is sometimes used in place of sexual orientation, this is not the meaning I am ascribing to the word. Sexuality, as I will use the term, denotes all aspects of sex: biological sex, sexual orientation, gender, gender identity, gender expression, and gender roles. I am using sexuality in this way because the different aspects of sexuality are connected, although they are certainly not the same. Much of the discussion about intersex involves biological aspects of sexuality. However, human sexuality is very significant to our humanity and it has been affected in multifaceted ways by the fall. I want to emphasize throughout that biological aspects of sexuality are not the only things that have changed due to the entrance of sin into the world. The distortion of sexuality affects every person on the planet, not merely people whose biology is ambiguous. This usage of the term sexuality, then, is intended to keep us all humble.

Overview of the Book

This book is a theology of intersex whose main contention is that the gospel of Jesus Christ is absolutely good news for persons with intersex traits. This book is written first of all for intersex persons themselves, in the hope that it will help intersex Christians to make sense of their own selves. Second, it is addressed to the church in order to shed some light in the darkness of ignorance. In showing how intersex fits with the gospel of Christ, the book aims to do three things: to provide intersex people with a new perspective about intersex, not based in shame but in the love of God; to open the church to a new position of inclusion and welcome for people whose sex is biologically ambiguous; and to inform a pastoral response to the issues that trouble intersex people. No doubt other groups of people would benefit from reading these pages. Medical professionals might see intersex differently. People in the LGBTQI community might find that God is not against the biologically unusual. And counselors may find value for their intersex clients within these pages.

The book begins with some basics about intersex. Chapter 1 explores the nature of intersex, including major intersex conditions. It outlines the biology of sexual development and present theories about how people become intersex. Then medical views on intersex and the diversity of intersex persons' views on the condition are contrasted. Issues for intersex persons include the significance of gender, gender dysphoria, identity, and shame. Parents also experience difficulties associated with the birth of an intersex child. There is a cultural agenda to use intersex to transform or abolish gender. Finally, the difference between intersex, transgender, and homosexuality is emphasized.

Chapter 2 considers what it means to be a created being. Humans are embodied creatures, fearfully and wonderfully made in the image of God. Human bodies and human sexuality are declared good by the God who created us in love. This is true for all bodies: female, male, or intersex. Humans were originally made as male and female. But sin has affected both our physical being and our sexuality. This fact does not separate intersex people from the goodness of bodies or the image of God, but it does imply that intersex is *one* result of the fall. However, there is real hope for those whose sexuality is unusual. In fact, God has used the sexually unusual to bring about his redemptive purposes.

Sex and gender are the subjects of chapter 3. Although there are injustices associated with gender roles and expectations, evangelical Christians cannot endorse the abolition of gender. Humans are made male and female to reflect the distinctions and unity within God. Male and female joining together in marriage also points us forward to the destiny of humanity as the bride of Christ. Although as Christians we cannot deny the importance of male and female, there is truly a positive place for intersex people in God's kingdom. Intersex people can become disciples of Christ and use their unique gifts within the church. The final part of the chapter considers intersex and gender through three lenses: the positive status of the eunuch in the New Testament, the positive value of singleness in the kingdom of God, and the intersex person and non-marital relationships.

Chapter 4 explores the life and ministry of Jesus. Jesus chose to experience the difficulties of human life, including many things that parallel the lives of intersex people. His lowly status, perilous childhood, and mockery over the manner of his birth have much in common with intersex lives. Therefore, he can truly empathize with intersex pain. Although there is no record an intersex person in the Gospels, the ministry of Jesus shows how he would treat the ambiguously sexed. The unclean were offered love and respect, not disgust. No one was excluded based on his or her biology, or blamed because of birth defects. Jesus ate with outcasts and gave grace to the struggling. The challenge to the church is to act in the same way.

In chapter 5 the impact of Jesus' death on intersex is considered. The betrayal and persecution of Jesus means that he knows what these experiences are like for intersex people. In his crucifixion, Jesus bore the shame of intersex, by being dehumanized and having his nakedness forcibly exposed to public display. He won forgiveness for sinners and calls intersex people to forgive those who have wronged them. Jesus lost his identity as Son of God in order to give intersex people a new identity as children of God. The death of Christ opened up new access to God in worship, overriding the Old Testament restrictions related to biology. Finally, in dying for his bride, Jesus enables the final fulfillment of sex and gender and the healing of broken sexuality.

Chapter 6 considers the resurrection of the dead and implications of the resurrection for the present. The resurrection body will be gloriously transformed and yet connected with the present body. Sexuality will be radically transformed for all people, however sexed. Intersex will not remain in the resurrection, but intersex believers can be sure that their identity is secure in Christ. The future resurrection also impacts life now. It affirms the goodness of bodies. This will help parents and intersex persons make decisions about surgery and other medical treatment. The resurrection provides hope, which can sustain intersex people through difficult situations that cannot be changed in the present. The resurrection is also the basis for Christian ethics, centered on love and empowered by the Spirit.

The final chapter is a call for the church to be inclusive of people who are "other," particularly intersex persons. Difference was given by God as a gift, but sin and false ideas about God prevent us from embracing the other. However, the work of Christ has created one church, thus including the other in his body. God displays his glory and manifests his presence in the weakness of intersex. Indeed, the church needs intersex people in order to be complete as the bride of Christ.

The love of God expressed in the gospel towards intersex people must be lived out in the church through acceptance, love, justice, and inclusion. This is the only truly evangelical response to intersex.

1

Intersex

Introduction

BEFORE EMBARKING ON THIS project I was largely ignorant about intersex conditions; I barely recognized the word "intersex." Ignorance is apparently widespread. Possibly this stems from the fact that intersex is rare and has been hidden for a long time at the behest of the medical profession. A recent discussion with a general practitioner has convinced me that even medical personnel may still be uninformed about intersex. Many people confuse intersex with transgender and yet these are completely different matters. Most likely many readers of this book will be similarly ignorant and confused.

Ignorance is a real problem because ignorance breeds fear. What we do not understand we avoid or despise. But intersex people are people for whom Christ died. Prejudice, fear, avoidance, and hatred are not appropriate Christian responses to people with ambiguous biology. The reason that this chapter is included is to dispel some of the ignorance. When the mystery and confusion are gone, it will be easier for the church to act in an inclusive way towards people with ambiguous biology, and this in turn may allow intersex people to be open about their own biology.

The theological discussions of the following chapters cannot be properly understood without first grasping what intersex is and what the issues are. For this reason the first chapter is dedicated to exploring the biological, psychological, cultural, and social issues surrounding intersex.

It begins with a basic definition, terminology, and numbers. Intersex is not one condition, but results from a number of different conditions. The major ones are described along with some other conditions that are sometimes grouped with intersex.

In trying to understand intersex it also helps to know about how people become male or female. Not everyone has the same opinions about intersex and how to respond to it. There is a substantial divide between medical professionals and intersex advocates.

A number of issues surround intersex, first for the intersex person and the parents, and then for the culture at large. Gender is something very important to social interaction. But the ambiguity of intersex bodies can take its toll on personal identity. Medical interventions and secrecy have been particularly damaging for many people. The damage is in some measure physical, but more significantly shame pervades intersex. Shame impacts both intersex people and parents.

Christians need to be aware that intersex people are being used as a "proof" that gender is fluid. They also need to be clear that intersex is biology not sin; homosexuality and transgender should not be confused with intersex.

What is Intersex?

Definition and Terminology

> Intersexuality is the biological condition of being "in between" male and female. There are many different kinds of intersexuality, and many different degrees of each. Sometimes intersexuality can be recognized by a mere visual inspection, such as when it is manifest in ambiguous genitalia. In other cases, genitalia will appear to be typically male or typically female, but will be discordant with the "sex" of the chromosome, the gonads, or both.[1]

Intersex is basically ambiguous sexual biology. The intersex person is somewhere between male and female, without being definitively either.

Intersex is a relatively recent term, replacing the old terms hermaphrodite and pseudo-hermaphrodite. The medical establishment now uses the terms disorders of sex development (DSD) or disorders of sex

---

1. Sytsma, "Introduction," xvii.

differentiation (DSD).[2] DSD terminology is, however, not without its detractors. The United Kingdom Intersex Association (UKIA) objects to the term disorders of sex development, because disorder implies that intersex is pathological, that is, a disease. Instead they consider intersex a "natural variation in human development." In the view of UKIA, intersex is a neutral term that can be used without implying disease or stigma.[3]

What terminology am I going to use? Someone has said that we do not so much need new terminology as we need new people. The nature of humanity is such that whatever terminology we adopt as neutral can over time become pejorative. It is not my intention to cause offense and increase the stigmatization of marginalized people. Unfortunately, there is no one choice of terminology that will satisfy all parties. I have chosen to use either intersex, DSD, or ambiguously sexed throughout. Occasionally there will be mention of hermaphrodites and pseudo-hermaphrodites, since these terms are frequently found in the literature. No negative connotations are intended to be attached to these terms.

How Many Intersex People Are There?

How many people live with an intersex condition? The answer depends on what conditions are included within the term, so estimates vary enormously. Some scholars have estimated that 1.7 percent of live births are intersex. However, this estimate is based on including a condition that does not often result in intersex traits. Without it, the estimate would be 0.228 percent of live births. If hypospadias (a malformation of the penis) and undescended testes are included then the estimate would be over 2 percent.[4] Others are much more conservative in their estimates. Leonard Sax believes that 0.018 percent of live births is far closer to the mark. If Sax's estimate is correct then approximately 50,000 Americans are intersex.[5] Let's put these numbers in context. Even the most conservative estimates show that the number of intersex infants born is about the same as the number of infants with hemophilia, or about half the number of people born with cystic fibrosis. Most people are familiar with these diseases. Yet intersex remains something hidden.

2. Barbaro, Wedell, and Nordenström, "Disorders of Sex Development," 119.
3. "Why Not 'Disorders of Sex Development'?"
4. Blackless et al., "How Sexually Dimorphic Are We?," 161.
5. Sax, "How Common is Intersex?"

Why does it matter how many people are intersex, especially since it is obviously not very many? In some sense it makes no difference. It should make no difference to whether or not people are loved, treated with dignity, and given support by others. However, there is a real association between rarity and freakishness. Parents cannot find support when they are told by doctors that no one else has gone through the same experience. If an intersex person feels like a freak, he or she may remain isolated and ashamed to ask for help or support. So a lack of information about frequency can perpetuate secrecy and shame.[6]

Major Intersex Conditions

There are a large number of conditions that result in ambiguous biological sex. In some cases the person has male chromosomes (XY) but female appearance. In other cases this is reversed. The intersex trait may or may not be obvious at birth; sometimes people do not find out until puberty or later. What they all have in common is the ambiguity. It is not realistic or needful to explore every possible intersex condition here. My purpose is to provide enough information so that the biological facts come into the open. Since there is a degree of disagreement about what constitutes an intersex condition and what does not, I will first explore conditions that are generally accepted as intersex and then outline some conditions that are sometimes grouped under the intersex umbrella.

Congenital adrenal hyperplasia (CAH) is the most common reason for intersex traits in genetic females. CAH can occur in genetic males, but in boys it not considered an intersex condition. It results from a malfunctioning adrenal gland, which produces too little cortisol and too many androgens. Classic CAH is dangerous, because it causes vomiting and dehydration, and may cause death if not treated. It is, therefore, one of the few intersex conditions that is life-threatening. CAH is an inherited condition, requiring the gene to be passed down by both parents. Approximately one in 13,000 babies is born with classic CAH.[7]

Girls born with CAH have external genitalia that are virilized, that is, they look like that of a boy. But the internal reproductive organs are female

6. Fausto-Sterling, *Sexing the Body*, 51; Herndon, "Do Frequency Rates Matter?"

7. Fausto-Sterling, *Sexing the Body*, 53; Muram and Dewhurst, "Inheritance of Intersex Disorders," 121; Preves, *Intersex and Identity*, 27; Warne et al., "Hormones and Me," 4, 8.

and functional. The degree of virilization of the genitals varies from simply having a slightly enlarged clitoris, to a clitoris that looks like a small penis, to the extreme case of looking completely masculine. Although these girls have a scrotum, they do not have testes. CAH results in other problems in later life, including early puberty, excessively masculine features, and premature bone growth that makes adults with CAH quite short. Women living with CAH need to take medication for their whole lives to prevent the appearance of male secondary sex characteristics.[8]

One CAH woman has described the traumatizing results of failing to take her hormone pills during adolescence. Her desire was to appear feminine and obviously so. However, when as a teenager she was told to keep taking her medications, the warning was that without them she would die. Since teenagers are rarely afraid of death, she gave up taking the drugs. Now she describes herself as "a freak caught somewhere between male and female." Her experience of her own body is so negative that she laments not taking the medications. "There are worse things in life than death and one of them is looking like this."[9]

Since the 1950s the recommendation of medical professionals has been to assign female sex to genetically female CAH babies, because CAH women are fertile. Feminizing surgery is often carried out to ensure a female gender identity, and this is effective in 90 percent of cases. However, CAH females whose genitals are more masculinized are quite likely to express a male gender identity and are less likely to have children. Those raised as males are usually sexually attracted to women. Some are happy with a male gender identity and have strong libido, functional penis, and satisfactory sex as males. But some severely masculinized women with CAH who had feminizing surgery and were raised as females have poor body image, lack self-esteem, and have attempted suicide as a result of their female sex assignment.[10]

The second major intersex condition is androgen insensitivity syndrome (AIS), which affects somewhere between one in 20,000 and one

8. Feder, *Making Sense of Intersex*, 24; Preves, *Intersex and Identity*, 27; Reiner, "Prenatal Gender Imprinting and Medical Decision-Making," 156; Sytsma, "The Ethics of Using Dexamethasone to Prevent Virilization of Female Fetuses," 242.

9. Chimera's story quoted in Preves, *Intersex and Identity*, 27.

10. Farley, "Challenging the Disorders of Sex Development Dogma"; Kim and Kim, "Disorders of Sex Development," 4; Slijper et al., "Long-Term Psychological Evaluation of Intersex Children," 127; Sytsma, "Intersexuality, Cultural Influences, and Cultural Relativism," 265–66.

in 13,000 people. Like CAH it is inherited. AIS women have XY chromosomes but have external genitals that appear female. The external anatomy is not, however, the complete picture. AIS individuals have testes, which are usually undescended or partially descended. They have no ovaries, no cervix, and no uterus. If a vagina is present it will be short and incomplete. AIS is sometimes diagnosed at birth, because the undescended testes look like a hernia, but often it is not diagnosed until puberty when the girl fails to menstruate.[11]

AIS women lack androgen receptors, which bind with testosterone so that it can be used by cells. Therefore, the hormones that cause the development of male genitalia in genetic males cannot be used properly by the body. This also affects puberty. Instead of the testosterone producing masculine secondary sex characteristics at puberty, the body responds as if it were estrogen. Thus AIS women develop breasts, but have very little underarm hair and often no pubic hair, because androgens are necessary to produce these in adolescence. Many women with complete AIS are considered very beautiful by Western esthetic standards. Since on the surface girls with AIS look like other girls, the discovery of their genetic maleness, at puberty or later, can prove quite a shock.[12]

Nicky (born 1943) has complete androgen insensitivity syndrome (CAIS) and was raised as female. Her experiences of adolescence with an unusual body were difficult. Nicky found out at age sixteen that she had no uterus. She was unable to join in when other girls talked about having periods. In high school she was traumatized by being forced to shower together with other girls after physical education, knowing that others would see that she had no pubic hair. Later in life there were more difficulties. In her late twenties, Nicky felt that she was not a real woman and struggled with her infertility.[13]

With CAIS the infants look like girls and are always reared as females. Partial androgen insensitivity syndrome (PAIS), on the other hand, is both much rarer and more variable in its outcomes. People with PAIS are born with different degrees of insensitivity to androgens. Depending on the extent of the insensitivity, external genitalia can range from male genitals,

11. AISSGA, "Androgen Insensitivity Syndrome"; ISNA, "Androgen Insensitivity Syndrome (AIS)".

12. Callahan, *Between XX and XY*, 103–4; ISNA, "Androgen Insensitivity Syndrome (AIS)"; Preves, *Intersex and Identity*, 28.

13. Callahan, *Between XX and XY*, 106, 108.

through ambiguous genitals, to female genitals at the complete AIS end of the spectrum.[14]

Five-alpha reductase deficiency (5-ARD) is a form of AIS which is very prevalent in certain pockets of the world—the Dominican Republic, Papua New Guinea, Mexico, Brazil, and the Middle East. The 5-alpha reductase enzyme functions in the body to convert testosterone to dihydrotestosterone (DHT), which is necessary for a developing fetus to form male genitalia. Without this enzyme, genetically male infants are born with predominantly female-looking genitals and undescended testes. However, at puberty these children develop male external genitalia. Infants with 5-ARD are generally raised as girls and transition to male gender identity in adolescence.[15]

Although classic CAH and AIS are the most common intersex conditions, there are many others. People with mosaicism have different chromosomes in different parts of the body. Some people are completely male, some completely female, and others have ambiguous genitalia. A chimera has different chromosomes in different cells of the body. While mosaics develop from one zygote, chimeras develop from two. A whole spectrum of potential maleness or femaleness can develop from a chimeric person.[16] True hermaphrodites are extremely rare, occurring in approximately one in 83,000 live births. To qualify as a true human hermaphrodite a person must have some ovarian tissue and some testicular tissue. Most are raised as male, because testosterone produces a phallus—either a penis or a very large clitoris. Most also have a vagina and a uterus. Most have at least one testicle. At puberty most grow breasts and menstruate and often ovulate; some who are chromosomally female have given birth to children. However, very few produce sperm.[17]

14. AISSG, "What Is AIS?"; Blackless et al., "How Sexually Dimorphic Are We?," 153; Slijper et al., "Long-Term Psychological Evaluation of Intersex Children," 126.

15. Blackless et al., "How Sexually Dimorphic Are We?," 153.; Reiner, "Prenatal Gender Imprinting and Medical Decision-Making," 157; Zucker, "Gender Identity and Intersexuality," 169–70.

16. Callahan, Between XX and XY, 67; "45,X/46,XY Mixed Gonadal Dysgenesis".

17. Callahan, Between XX and XY, 76–77; Wilson and Reiner, "Management of Intersex," 122.

Other Conditions

The conditions described above are the most commonly occurring, uncontested intersex variations. A few other conditions are sometimes included as intersex: Klinefelter syndrome, Turner syndrome, and hypospadias. There is some debate about whether these should be included, because the ambiguity of sex is not always apparent. However, some people with these conditions experience similar issues to those who have clear intersex conditions. It matters less about the label of intersex or not intersex, and more about the confusion, prejudice, and debatable medical management that has been the experience of many people with these conditions.

Klinefelter syndrome is the most common syndrome resulting from an unusual number of chromosomes, affecting approximately one in a thousand men. Most humans have forty six chromosomes, including one X and one Y chromosome (46,XY). Men with Klinefelter syndrome have 47,XXY chromosomes or other variations of more than one X, and sometimes more than one Y chromosome. Symptoms commonly include learning disabilities, increased height and limb length, little body or facial hair, and small testicles. Klinefelter boys are frequently passive, shy, and immature. Many men with Klinefelter are sterile. About half of boys with Klinefelter develop breast tissue at puberty and commonly have female body fat distribution.[18]

One man with Klinefelter describes his experience of difference:

> It wasn't until I was 29 years old that a label was put on my physical differences, differences that I never quite understood. I had large nipples on smallish breasts, peanut-size testicles, and cellulite-type hairless fatty tissue over most of my body. . . . The medical journals called my condition "feminized male." I had always felt caught between the sexes without knowing why.[19]

Turner syndrome names the chromosomal anomaly in which a person has only forty five chromosomes, including only one X. Most fetuses with Turner syndrome do not survive. Girls with Turner syndrome are externally clearly female and the vagina and uterus are generally properly developed. However, in place of ovaries there are streak gonads, which do not produce the appropriate level of hormones. These do not signal puberty in

---

18. Callahan, *Between XX and XY*, 62–63.; Fausto-Sterling, *Sexing the Body*, 53.; NSW Government Health: Centre for Genetic Education, "Fact Sheet 31," 1.; Woodward and Patwardhan, "Disorders of Sex Development," 401.

19. Cameron, "Caught Between," 91.

Turner girls, so they do not develop breasts or menstruate. Turner women are sterile and are usually very short.[20]

One other condition which is not correctly an intersex condition, but which presents some similar issues, is hypospadias. A hypospadias is a malformation of the penis in which the urethral opening is not at the tip, but rather on the shaft or base. Surgeries on hypospadias are common and done so that the boy can stand while urinating. The surgeries frequently result in urinary tract infections and difficult and painful urination.[21] Numerous surgery horror stories can be found. Howard Devore had twelve surgeries. He writes:

> The promise that you will be able to pee standing up is just plain false. . . . I believe that they know that, but is seems that genital appearance and the promise of normalcy are more important to young parents than a clear-headed acceptance of reality.[22]

Even more shocking is the cry from a mother of a young boy, trying to find a therapist to deal with gender dysphoria resulting from surgical trauma. After two hypospadias surgeries, her five-year-old son began to say, "I hate my dong [penis], I wish I was born a girl."[23]

In sum, intersex is not one condition but an umbrella term, which covers a variety of conditions, each of which results in some degree of ambiguity of biological sex. Sometimes that is obvious at birth and sometimes it does not become evident until later. No one chooses to be intersex. Intersex people are simply born with a biological condition. Intersex has social, cultural, and psychological implications, but these are the result of reactions to the biological facts of intersex. These implications are explored later in the chapter. But before considering these, I offer some biological explanations for the existence of intersex humans.

Male and Female

To understand how a person is born intersex, we need to know something about sex, starting with what sex is. It is helpful to know how the normal

20. Callahan, *Between XX and XY*, 63–65; Preves, *Intersex and Identity*, 30.

21. Preves, *Intersex and Identity*, 31.

22. Devore, "Growing Up in the Surgical Maelstrom," 81.

23. Chase, "Surgical Progress Is Not the Answer to Intersexuality," 152.

process of sex development takes place. With this foundation it is possible to theorize about how sexual development can go wrong.

What is Sex?

The first thing we must understand is what we mean by sex. Sex has many facets and is therefore not as simple as it may appear at first blush. Sex can include the following: "genetic sex, chromosomal sex, gonadal sex, phenotypic sex (internal), phenotypic sex (external), sex of rearing, gender (childhood), gender (adulthood)."[24] Usually these different aspects of sex develop in a corresponding way in a particular individual, so that a person is entirely female or entirely male. But this is not always the case.[25]

Genetic sex and chromosomal sex are often the same, but in some cases may be different. The genetic sex is determined by whether or not there is a gene sequence that switches on the development of testes. The chromosomal sex is decided by whether the person has a Y chromosome or not. Gonadal sex is determined by the gonads; males have testes and females have ovaries. In the womb, hormones secreted by the embryo's gonads bring on the development on the internal sex structures of the infant. Secretion of hormones then brings about the internal phenotype sex. External phenotype sex refers to the external genitalia. Sex of rearing is the decision of the parents to raise the child as either a girl or a boy. Parents behave differently toward a boy or a girl. Gender, on the other hand, is less about anatomy and more about "the individual's psychological and emotional identification as male or female."[26]

Since intersex is foremost about biological sex, a scientific definition of male and female is helpful. Here is one such definition:

> [T]he typical male [is] someone with an XY chromosomal composition, and testes located within the scrotal sac. The testes produce sperm which, via the vas deferens, may be transported to the urethra and ejaculated outside the body. Penis length at birth ranges from 2.5 to 4.5 cm; an idealized penis has a completely enclosed urethra which opens at the tip of the glans. During fetal development, the testes produce the Mullerian inhibiting factor, testosterone, and dihydrotestosterone, while juvenile testicular

24. Wilson and Reiner, "Management of Intersex," 120.
25. Kemp, "The Role of Genes and Hormones in Sexual Differentiation," 1.
26. Wilson and Reiner, "Management of Intersex," 120–21.

activity ensures a masculinizing puberty. The typical female has two X chromosomes, functional ovaries which ensure a feminizing puberty, oviducts connecting to a uterus, cervix and vaginal canal, inner and outer vaginal lips, and a clitoris, which at birth ranges in size from 0.20 to 0.85 cm.[27]

Given this biological definition, we now consider how an individual becomes biologically male or female. The answer may help shed light on how an individual does not become clearly one or the other.

Normal and Abnormal Sexual Development

Most people are born with twenty-three pairs of chromosomes. The twenty-third pair is called the sex chromosomes and is generally either XX or XY. Normal development results in either a 46,XX girl or a 46,XY boy.[28] In the womb, fetuses are not sexually differentiated in appearance until week six of gestation. At this stage the fetus has both Müllerian (female) and Wolffian (male) ducts. The gonads have potential to develop into either testis or ovary and begin to differentiate based on the chromosomes of the fetus. A Y chromosome results in the formation of testes and male sexual development. When testosterone is released, Wolffian ducts develop into internal male genitalia. External male genitalia develop when testosterone is converted into dihydrotestosterone (DHT). Anti-Müllerian hormone (AMH) causes female structures to shrink. In the case of a fetus with two X chromosomes, the ovaries begin to develop in week seven of gestation. Because the fetus does not have any AMH, the Müllerian ducts are able to develop into internal female genitalia. The lack of testosterone allows the Wolffian ducts to shrink. The fetus begins life with undifferentiated external genitalia. But when the male fetus produces DHT, the penis and urethra are formed and genital folds become the scrotum. Without DHT, the female fetus will develop a clitoris, labia minora, and labia majora.[29]

There are two general ways in which a disorder of sex development may occur: as a result of genetic aberrations or as a result of hormonal production or reception. The most likely reasons for genetic abnormalities are a loss of one of the copies of the gene or a duplication of the genes. If gonads have developed normally, the problem is likely a hormonal defect

27. Blackless et al., "How Sexually Dimorphic Are We?," 152.

28. Preves, *Intersex and Identity*, 23.

29. Barbaro, Wedell, and Nordenström, "Disorders of Sex Development," 119–20.

affecting the external genitalia. There are some factors that influence the mis-production of hormones and potentially cause intersex. In wildlife and laboratory animal studies, there is evidence that when pregnant animals are exposed to chemicals that have an effect on endocrine systems, intersex off-spring are born. It is possible that this may also cause intersex in humans.[30]

Science provides us with some explanations of intersex as a biological phenomenon based on disruptions or distortions to normal fetal development. Yet human sexuality is such a fundamental part of human existence that biological discussions of intersex can never be enough. There are deeper issues surrounding intersex than the mere objective facts of biology. The next section begins the discussion of these deeper issues by considering how the medical profession and others have responded to intersex.

Perspectives on Intersex

The response of the medical profession to the fact of intersex is different to that of intersex advocates. The medical treatment of DSD has undergone quite a change over the last two decades, but medicine generally considers DSD a pathology that required surgical intervention. This perspective has been criticized by intersex advocates, who suggest instead that gender should be socially assigned. However, intersex persons are not a homogenous group. There are differing views about the condition and about the place of surgery.

Medical Advice, Responses, and Changing Paradigms

Possibly the matter that has been most criticized by intersex individuals is the medical paradigm for dealing with intersex births. Although in recent years the medical paradigm is in the process of change, a large number of intersex persons have had surgery and hormone therapy with which they are unhappy. In this section I will explore what the medical establishment thinks about intersex and how they advise parents of an intersex child.

In 1969 doctors John Dewhurst and Ronald Gordon expressed what many in the medical profession no doubt thought about intersex births:

---

30. Ibid., 121, 23, 26; Juul, Main, and Shakkebaek, "Disorders of Sex Development," 505; Hughes, "Disorders of Sex Development," 128–29.

That a newborn should have a deformity [affecting] so fundamental an issue as the very sex of the child . . . is a tragic event
which immediately conjures up visions of a hopeless psychological
misfit doomed always to live as a sexual freak in loneliness and
frustration.[31]

In 1984 the advice of Dewhurst and Muram to doctors on how to deal
with intersex newborns appears to be based on the same sense of tragedy:

A newborn infant with genital ambiguity represents *an acute medical emergency*. With appropriate management the child may have
a happy, well adjusted life and may even be fertile. However, with
inappropriate or no management the child may be considered a
"freak" and therefore suffer loneliness and frustration. The sex of
the child should be "assigned" as soon as possible. Regardless of
the complexity of doing so, the choice will often be the main de-
terminant of the outcome for the child and family.[32]

For a long time medical professionals believed surgery was the solution to the "acute medical emergency" described by Dewhurst and Muram.
Surgery as a solution to intersex was adopted in earnest in the 1950s. At
that time, John Money, a psychologist at Johns Hopkins, believed that to
develop a stable gender identity people require genitals that match the sex
assignment. According to Money, children are gender neutral at birth and
nurture produces a gender identity. In accord with this idea, doctors can
then underpin the sexual identity of the intersex infant through surgery
and hormone therapy. This policy was termed the "optimal gender policy."
Boys whose penises were considered undersized were surgically altered
and raised as girls. On the other hand, infants born with enlarged clitorises
were likely to have these surgically removed or hidden. The focus was on
cosmetic appearance of genitals, rather than any consideration of sexual
function. As a consequence, many people who had genital surgery as infants lack sexual sensation.[33]

In recent years there has been a shift in the medical practices surrounding intersex births. The optimal gender policy fails to grasp the

31. Dewhurst and Gordon, *The Intersexual Disorders*, 52, quoted in Holmes, *Intersex*,
140.

32. Muram and Dewhurst, "Inheritance of Intersex Disorders," 124. Italics mine.

33. Chase, "Surgical Progress Is Not the Answer to Intersexuality," 148; Creighton
and Minto, "Managing Intersex," 1265; Crouch, "Betwixt and Between," 32; Dreger, "A
History of Intersex," 10–12; Preves, *Intersex and Identity*, 56–57; Wiesmann et al., "Ethical Principles," 671–72.

complexity of gender identity, produces trauma through surgery, and does not allow for the consent of the child. Therefore, medicine has adopted a new paradigm called the "full consent policy." It believes gender must be assigned at birth, since no person can be a productive part of society without a gender. But the old approach was far too paternalistic. Now a team of doctors work together to determine the most appropriate sex, based on analysis of what has been most successful for other intersex people. Surgery on genitals is not considered urgent. However, there are still many people who are deeply concerned about the stigma attached to DSD and perform surgery on infants despite the new paradigm.[34]

There are some potentially valid medical reasons for some of the surgeries routinely performed on intersex persons. In the main, this has to do with the increased risks of certain cancers in XY women.[35] However, the propensity of medical personnel to recommend surgery for intersex births has produced a lot of literature by intersex advocates, who want alternatives to surgical intervention. Cheryl Chase, founder of the now defunct Intersex Society of North America (ISNA), insists that surgery does not result in normal genitals. Anyone who has looked at the photos of the supposed surgical successes would know this to be true. Instead of producing normal genitals, surgery sends a message that intersex genitals are offensive. The surgical procedures on intersex children communicate loudly that children with unusual genitals are unlovable as they are and must be fixed.[36]

Instead of surgery in infancy, many suggest social means of assigning gender to a child. In general social situations people do not consider genitals in order to determine the gender of an individual. Instead we rely on social cues such as names, clothing, hair styles, gestures, body shape, and interests. So a person would not need genital surgery to be recognized as a particular gender. But intersex groups are clear that the gender assignment should be provisional. Counseling is advocated to allow the child to think through her or his gender identity. Then, if as an adult that person desires surgery, it should be made available.[37]

---

34. Callahan, *Between XX and XY*, 116–20.; Wiesmann et al., "Ethical Principles," 671–72.

35. Ismail and Creighton, "Surgery for Intersex," 62.

36. Chase, "Surgical Progress Is Not the Answer to Intersexuality," 155; Preves, *Intersex and Identity*, 32.

37. Ozar, "Towards a More Inclusive Conception of Gender-Diversity for Intersex Advocacy and Ethics," 17.

Other Intersex Views

It would, however, be a mistake to assume that all intersex persons hold the same opinions about intersex issues, including decisions about surgery. Since many people who are intersex have no interest in making their lives public, we cannot know whether or not they are in agreement with intersex advocacy groups. Even those whose opinions are made known are not uniform in their perspectives on being intersex.

To begin, there are differing views on surgery. Many intersex individuals in the West are opposed to genital surgery. However, this view is not necessarily shared by intersex persons in other countries. In poorer countries, genital surgery on intersex persons is usually carried out late in childhood, during adolescence, or sometimes on adults. This contrasts with the situation in Australia, where surgery has been routinely performed on infants between four and six weeks old. Some intersex people in those poorer countries would have preferred surgery early in life. They want no memory of the surgery. They were also concerned about bullying, a lack of self-assurance because of being different, confusion about gender, and psychological pain that results from being compared to "normal" children.[38] This demonstrates that Western views about surgery are not the only view. Indeed there are studies that indicate that many intersex persons in the West are not unhappy with their surgeries.[39]

Surgery is not the only issue on which intersex persons differ. Some, like Cheryl Chase, are proud of being intersex:

> My body is not female; it is intersexed. Nonconsensual surgery cannot erase intersexuality and produce whole males and females; it produces emotionally abused and sexually dysfunctional intersexuals. If I label my post-surgical body female, I ascribe to the surgeons the power to create a woman by removing body parts; I accede to their agenda of "woman as lack"; I collaborate in the prohibition of my intersexual identity.[40]

We must not conclude that there is only one kind of intersex experience. There is room for a variety of experiences and a variety of opinions among intersex persons. Some intersex persons, for instance, have aligned themselves with LGBTQ advocates. But others do not want to identify

38. Warne and Bhatia, "Intersex, East and West," 186, 90–91.

39. Ibid., 202.; Holmes, *Intersex*, 16.

40. Chase, "Affronting Reason," 214, quoted in Holmes, *Intersex*, 108–9.

with LGBTQ issues, since they say that being intersex does not challenge their maleness or femaleness or their heterosexuality. Some want to have the option to tick a box for intersex or other on official forms. But this idea is frightening for others.[41] Emi Koyama, director of Intersex Initiative, comments:

> [The] vast majority of people born with intersex conditions live normally as a woman or a man, and do not view themselves as a member of a different gender/sex category. Most people born with intersex conditions do not think "intersex" to be who they are; it is simply a medical condition, or a lived history of medicalization. Most people with intersex conditions would answer "no" if they are asked "are you intersex?" . . . To list "intersex" along with "male" and "female" gives the false impression that one cannot be male or female if she or he has an intersex condition. This hurts people with intersex conditions who identify as male or female, and mis-educates the general public.[42]

It is true that the medical profession has begun to change its response to intersex. It is also true that not every intersex person has the same view or the same experience. Some may even be content with the surgery performed on them in infancy. However, for many intersex people, medical management has been a negative experience.

Issues for Intersex Persons

The physical aspects of medical management of intersex are by no means the only issues of consequence to intersex people. Gender is such a significant aspect of identity and human interaction that having ambiguous biology can result in gender dysphoria and identity confusion. But it is the shame that seems to be the major issue for people with intersex traits. The surgery sends a message that the person is not acceptable without it. Repeated medical examinations of genitals by multiple doctors and being forced to be photographed naked produce significant shame. Some intersex individuals have labeled the medical treatment as sexual abuse. The shame is exacerbated by the secrecy that has surrounded intersex. All these issues deserve attention and are consequently explored below.

41. Grabham, "Citizen Bodies," 33–34, 41.
42. Koyama, "What is Wrong with 'Male, Female, Intersex.'"

The Importance of Gender

Gender is vitally important to a person's existence within society. With few exceptions, society is ordered according to an individual's gender. Men and women fulfill different roles in society. In general, men and women wear different clothing, have different behavioral expectations, use language differently, and pursue different interests. Schooling, employment, relationships, sporting competitions, religious obligations and roles, and requirements in law are all structured around gender. It is almost impossible to participate in society without acting within the strictures of gender. Every social interaction is predicated on knowledge of the other's gender. People usually prefer to be clearly male or clearly female and obviously so to others, because individual identity cannot be divorced from sex and gender.[43]

When the category of gender is removed from any social interaction, people become very disturbed and find it difficult to know how to behave towards the other person.[44] In 2010, the *Guardian* reported on a Swedish couple who refused to reveal the sex of their child. The child is never referred to by gendered pronouns, but only by the name Pop. The parents claim that Pop will grow up without stereotyped gender expectations.[45] The responses to the Swedish parents' decision were in some cases hysterical and even abusive.[46] Jane McCredie comments about the significance of this reaction:

> One thing the reaction to Pop's story does highlight is how important sex and gender are to us as ways of classifying people and how uncomfortable we feel when those categories are taken away from us. For all the increased freedom of our Western societies, the greater flexibility about both male and female roles at and outside work, we still seek those reference points of male and female when we think about or interact with other people.[47]

43. Preves, *Intersex and Identity*, 5, 14–15; Warne and Bhatia, "Intersex, East and West," 184.

44. Ozar, "Towards a More Inclusive Conception of Gender-Diversity for Intersex Advocacy and Ethics," 27.

45. Dowling, "The Swedish Parents Who Are Keeping Their Baby's Gender a Secret."

46. McCredie, *Making Girls and Boys*, 180.

47. Ibid., 180–81.

Gender Dysphoria

Since gender is such an important part of social interaction, it is difficult for those who are uncomfortable with the gender they have been assigned. The American Psychiatric Association (APA) defines gender dysphoria as a "marked incongruence between one's experienced/expressed gender and assigned gender, of at least 6 months' duration."[48] People with gender dysphoria are not just occasionally unhappy with birth gender, but experience deep distress. The APA estimates that somewhere between one in 7,000 and one in 50,000 adults have gender dysphoria.[49]

Gender dysphoria is an issue for many intersex people. The hope of doctors and parents is that the child will grow into an adult who is happy with the gender assigned, but this is often not the case. Different types of DSD have different rates of gender dysphoria. Studies suggest that for persons with CAH, the rate of gender dysphoria is about one in ten. For other conditions rates vary enormously, from one in eight for true hermaphrodites, one in five for PAIS, to sixty three in one hundred for people with 5-ARD.[50] It is clear from these statistics that the rates of gender dysphoria in intersex individuals are far higher than that of the general population. Therefore, a large percentage of people with intersex conditions will be very distressed in their assigned gender.

Living in a body that is perceived to have been assigned the wrong sex by doctors is emotionally difficult. One intersex individual exclaims, "While doctors still continue to spread the belief that assigning a child as a girl or boy is extremely crucial to their well-being, for those of us who they chose [the] wrong [sex for], our lives are just tortured."[51] This distress is almost impossible to escape since the body is always present and its significance to identity cannot be underplayed. According to Morgan Holmes, "'The body' is not simply an anatomical structure, but a symbolic one that conveys meaning, most fundamentally about one's gender, which is the meaning through which all others become possible."[52] Therefore, bodies, and what has been done to them, are tied up with the mental health of the person.

48. *DSM–5*, 452, cited in Yarhouse, *Understanding Gender Dysphoria*, 85–86.

49. Yarhouse, *Understanding Gender Dysphoria*, 85, 92.

50. Furtado et al., "Gender Dysphoria Associated with Disorders of Sex Development," 621–22.

51. Quoted in Callahan, *Between XX and XY*, 74.

52. Holmes, *Intersex*, 17. Italics original.

Identity

Many intersex persons report identity issues because of their intersex status. Most people do not even think about whether they are male or female, but intersex people do. Cheryl Chase asks, "Who am I?" She insists that she is both a woman and intersex, thus not female. Her identity as an intersex person is the result of her troubled gender. Chase looks female, but she cannot see herself as completely female. Women's clothes are the wrong shape for her female body and the size of her hands and feet. But men's clothes do not fit either, since she has curves like a woman. This poor fit is a constant reminder of what has been done to her body.[53]

For others, discovery of their intersex status later in life brings about an identity crisis. Sarah's identity was turned on its head when she discovered her XY chromosomes. She was no longer certain about who she is, even though nothing had really changed. She stills lives as a woman, since she has always known herself that way. And yet she began to ask, "What am I? Am I some kind of freak?"[54] Sexual attraction is also sometimes called into question. Melody began to wonder about her true sexual attraction when she was diagnosed with AIS. If she is actually male then should she be attracted to women? And yet she is not. So the idea that she might be homosexual entered in, only to be denied in the same breath.[55]

Other women with male chromosomes feel that they are in some way fraudulent as females. When Greta learned at age twenty-eight that she is XY, she questioned her femaleness. Greta says, "There was that [sense of], 'Oh my god, I'm XY. Oh my god, I'm really a guy.' I was a mess because I [felt like] a fraud. On the outside I'm a girl, [but] on the inside I'm a man."[56] Martha Coventry, a woman whose large clitoris was removed when she was six, fears exposure as not a real woman.

> When I was growing up, and well into adulthood, I used to have a waking nightmare that a squad of men in uniforms would arrive at my door, take me into the night, and execute me for not being a real woman. In my mind, they were always justified and I never raised my voice in protest.[57]

53. Chase, "Affronting Reason," 212, cited in Preves, *Intersex and Identity*, 90.

54. Preves, *Intersex and Identity*, 111.

55. Ibid., 112.

56. Ibid.

57. Coventry, "Finding the Words," 71.

Stigma and Shame

Although identity can clearly present issues for intersex persons, the most overwhelming issue is that of shame. The shame is connected to the surgical intervention, the negative way in which doctors and parents perceive intersex, repeated medical examinations, and the secrecy that surrounds the condition.

According to Alice Domurat Dreger:

> The thing that people with intersex suffer from the most is shame, it's not surgery. The surgeries are motivated by shame. So I think the bigger issue is people getting the message that [people with intersex] are not human, that they are not acceptable, that they are not loveable. That's a much bigger issue for everybody than the surgery is really. Because, what people who are anatomically different need the most is the message that they're human and acceptable and loveable.[58]

Early genital surgery is intended to prevent psychological problems in intersex children. Surgeons tell parents that surgery is urgently required and parents imbibe the idea that intersex is utterly shameful. So parents, wanting to keep children from the shame of being different, agree to the surgery. But the message that intersex is shameful is passed on to the children through the attitude of the parents. Thus the desire to erase difference through surgery does not prevent shame, but rather causes it. As a consequence, one study observed that 39 percent of children who had experienced intersex surgery had some kind of mental illness or psychological distress.[59] Surgery forces the intersex patient "to conclude that he or she remains unacceptable."[60]

The experience of shame in intersex individuals (and their parents) is itself not surprising given that the shame is present with the doctors themselves. Doctors report that an intersex birth is deeply traumatizing. One doctor comments:

> After stillbirth, genital anomaly is the most serious problem with a baby, as it threatens the whole fabric of the personality and the life

58. Quoted in Callahan, *Between XX and XY*, 155.

59. Howe, "Intersexuality: What Should Care Providers Do?," 214; Roen, "Clinical Intervention and Embodied Subjectivity," 33; Slijper et al., "Long-Term Psychological Evaluation of Intersex Children," 137.

60. Chase, "What Is the Agenda of the Intersex Patient Advocacy Movement?," 4.

of the person. The trauma of discovering a genital anomaly in the labour ward is great for both parents and doctor.[61]

Although many surgeons have begun to rethink surgical normalization of intersex infants, the disgust is still internalized. One particular doctor had a change of mind after taking some time studying bioethics.[62] After being formally interviewed, this particular doctor made an informal comment that betrayed his disgust with the genitals of intersex children. "Gesturing toward to the image [of an enlarged clitoris], he said, 'But then you have cases like that. What are you gonna do with *that*? I mean, you can't leave it like that.'"[63] Another pediatric urologist is quoted as saying, "Have you seen a baby with CAH? It's grotesque."[64] Chase insists that several surgeons have privately told her that taking the risk of compromising sexual function is always preferable to leaving an intersex child as is.[65]

Beyond the surgery, the repeated medical examinations are a powerful source of shame for many people. One woman with CAH describes it this way:

> The worst thing is being put in a prone position, half-naked, [and] told to spread your legs while five or six other people look in your crotch and probe. That definitely had a direct effect on my sexuality. It is embarrassing. It is shameful. It's painful. It's really painful.[66]

Sherri has CAIS. She remembers:

> I spent my adolescence filled with shame, though I was never told the true details of my diagnosis. My trauma was needlessly compounded by my doctor's stony silence while examining me, and his asking me to lie naked on an examining table so that teams of interns and residents could inspect my genitals. Such experiences themselves, far more than the true facts I later learned about the nature of AIS, instilled a sense of freakishness that I have only recently shaken.[67]

---

61. Hutson, "Cliteral Hypertrophy and Other Forms of Ambiguous Genitalia in the Labour Ward," quoted in Chase, "What Is the Agenda of the Intersex Patient Advocacy Movement?," 2.

62. Feder, *Making Sense of Intersex*, 68.

63. Ibid., 69.

64. Karkazis, *Fixing Sex*, 146.

65. Chase, "Surgical Progress Is Not the Answer to Intersexuality," 150.

66. Quoted in Preves, *Intersex and Identity*, 67.

67. Groveman, "The Hanukkah Bush," 26.

Jeannie writes:

> I was then placed on a table in a paper gown, feet in stirrups, probed and prodded with God knows what in my genital area, breasts examined, finger up my rectum, with three other interns and a nurse watching, asking questions as if I wasn't even there. Each intern touched my genitals. They took pictures. I was crying and nobody cared or stopped what they were doing. I begged them to stop and they just said "Oh that doesn't hurt that much" and "We'll be done in a minute honey."[68]

The secrecy surrounding intersex is very damaging and adds to the shame. Intersex persons have been told never to tell anyone else about their condition. Parents were told to keep quiet about the child's condition, to not even speak about it to family or friends.[69] Kiira Triea, a female pseudo-hermaphrodite, born in the mid-sixties and treated at Johns Hopkins, writes:

> I never uttered a peep about being intersexed to anyone, not a word. Ever. It was something that I seemed completely incapable of doing. It was unthinkable to even think about doing such a thing. Talking means death. Intersexed people don't talk.[70]

The conspiracy of silence includes some parents refusing to speak the truth to their adult children. Kailana, born with CAH, diagnosed with mosaicism, and told by doctors that he/she is a true hermaphrodite, talks about the secrecy that his/her parents still insist on. His/her parents will not admit to the intersex condition or the medical treatment that was given to him/her as a child. They refuse to discuss the topic at all. On occasion a word has been spoken and later denied.[71]

There are other results of secrecy. Secrecy leads to a lack of information, which in turn makes people imagine all sorts of horrors. The truth is easier to live with than lies.[72] Angela laments, "Although the doctors has [sic] claimed that knowing the truth would make me self-destructive, it was not knowing what had been done to me—and why—that made me want to

68. Jeanne's story quoted in Hester, "Intersex and the Rhetoric of Healing," 54.
69. Feder, *Making Sense of Intersex*, 43, 55, 73, 74, 90.
70. Quoted in Holmes, *Intersex*, 148–49.
71. Callahan, *Between XX and XY*, 70.
72. Groveman, "The Hanukkah Bush," 27.

die."[73] The silence can also lead to self-blame, as with one person with CAIS, who was raised as a boy. When she reached puberty she developed breasts. She wondered whether the breasts had grown because she had failed to act enough like a boy. Secrecy leaves intersex people without a voice, and without support. Doctors often told people that there was no one else with their condition. The result is isolation and a sense of freakishness.[74]

Many intersex people perceive the repeated medical exams of their genitals, and reconstructive surgery, as sexual abuse. Children are hugely vulnerable and intersex children even more so, because of the lack of control over their own bodies.[75] One woman described her vaginoplasty (surgical construction of a vagina) during her adolescence in this way: "I was forced to be surgically mutilated and medically raped at the age of fourteen. And that's exactly what I consider it."[76] Another woman compared her childhood sexual abuse with medical exams: "After I got to be a teenager, I could see a very clear parallel because you have somebody who is able to examine your body, and you don't have the right to say, 'No, you can't do this.'"[77]

For some people the shame has transformed into an intense self-hatred. Angela Morena had her large clitoris removed at age twelve. No one ever spoke about what was done.[78] She describes the fallout from this experience:

> Now 24, I've spent the last 10 years in a haze of disordered eating and occasional depression. My struggle with bulimia has been an all-consuming although mostly secret part of my life, and I now believe it represents my attempts to express the fear, shame, rage, and intense body-hatred that I have felt as a result of the—until now—unspeakable assault that I experienced under the guise of medical treatment.[79]

The biological facts of intersex can result in real issues for intersex people, including gender dysphoria and identity questions, but shame is above all the most significant problem. Unfortunately, these problems

73. Angela quoted in Hester, "Intersex and the Rhetoric of Healing," 54.

74. Chase, "Surgical Progress Is Not the Answer to Intersexuality," 149; Preves, *Intersex and Identity*, 74–76.

75. Preves, *Intersex and Identity*, 72.

76. Quoted in ibid., 73.

77. Ibid., 72.

78. Morena, "In Amerika They Call Us Hermaphrodites," 137–38.

79. Ibid., 138.

begin long before the child has any thoughts about his or her own body. Parents are impacted first of all.

Parents

Parents experience intense emotions and stress when an intersex infant is born. Some of the issues for parents are similar to those for intersex persons. These revolve around the identity of the child and shame. Parents experience shame, grief, loss of stability due to the liminality of the child's status, difficult medical decisions while isolated from family and friends, a general inability to cope emotionally with the birth of an unusual child, and relationship breakdown with their adult children due to decisions made for that child.

Intersex persons are not alone in feeling shame about intersex. The shame begins elsewhere. Since intersex challenges the general assumptions that Western society holds regarding sexuality, parents perceive the birth of an intersex child as an occasion for shame. Because of this, many parents are highly motivated to remove the source of their shame, in this case the ambiguous genitalia of the child. Many live in a state of perpetual denial, denying that the issue ever existed. Others have openly rejected the child. Parents are terrified to let friends and family know about their child's ambiguous sex, perhaps because physical difference is perceived as monstrosity. Shame adheres to the parents throughout their lives as they seek to hide the truth about their child's ambiguous status.[80]

Parents experience great stress upon the birth of a child with ambiguous genitalia. A child that cannot be fitted into the two-sex system lacks a status. Some feel that the child lacks a proper identity, and others that the child is not human. There is no language available to express the status of the child, so the child must be spoken of in negative terms, expressing what it is not. Even health professionals experience this absence of positive language categories. Floundering health professionals are a shock for parents, who believe that the doctor will know everything there is to know about the situation. The parents are left confused and disillusioned.[81]

---

80. Callahan, *Between XX and XY*, 70; Feder, *Making Sense of Intersex*, 63; Gough et al., "'They Did Not Have a Word,'" 500; Holmes, *Intersex*, 59; Slijper et al., "Long-Term Psychological Evaluation of Intersex Children," 132.

81. Gough et al., "'They Did Not Have a Word,'" 499–500.

Fear and grief are real issues for parents. Parents fear that without a clearly defined sex their child will be bullied by peers, be locked out of employment, and have no hope of forming relationships. There is anxiety about others seeing the child (or future adult) naked and thus having the ambiguity exposed.[82] Parents grieve the loss of the normal child they expected. Mary explains that when they discovered their daughter's PAIS, her husband, "just broke down and sobbed in my arms. That's when it impacted me the most. . . . There were a lot of tears, a lot of feeling bad for Jessica, knowing that she couldn't have children naturally."[83] Fathers in particular mourn the loss of a son that later turned out to be a daughter.[84]

After the diagnosis, the parents are faced with decisions about surgery. Although making medical decisions on behalf of their children is normal for parents, parents of an intersex child are placed in a very difficult position. They are conflicted between cultural expectations to produce a normally sexed child and the need to uphold the child's best interests.[85] An experiment conducted on two separate groups of college students illustrates this tension. In the first group, almost all indicated that they would not have wanted to have genital surgery if they were born with overlarge clitoris or micropenis. However, a different group of students—asked to imagine that they had a child with ambiguous genitalia—indicated that they would consent to cosmetic surgery for the child. Most parents cannot imagine themselves in the place of the child, because they are unable to disentangle themselves from the unconsciously held social expectations of normal.[86]

The story of true hermaphrodite Lisa May Stevens illustrates how some parents totally fail to cope with the intersex status of the child. Lisa May was named Michael at birth. But at age five or six his mother would call him Lisa May and force him to dress in girl's clothes. Since Michael did not want to dress as a girl, he would resist and his resistance was rewarded with physical violence until he complied. Michael's mother was happy and fun to be with when she dressed Michael as Lisa May, and Michael felt loved. Michael's father acted quite differently. He insisted that Michael must act like a boy, being tough and strong. Crying or tantrums were unacceptable

---

82. Ibid., 503; McCredie, *Making Girls and Boys*, 71.

83. Quoted in Feder, *Making Sense of Intersex*, 56.

84. Slijper et al., "Long-Term Psychological Evaluation of Intersex Children," 130.

85. Holmes, *Intersex*, 53.

86. Feder, *Making Sense of Intersex*, 44–47, 57.

behavior. Michael had to be careful to behave the "right" way with whatever parent he was with, because if he got it wrong he might get hurt.[87]

Ruby's story, on the other hand, illustrates what may happen to parents who do the best they are able. Ruby has two daughters with CAH. Both had surgeries. Getting appropriate medical care for her daughters was difficult, both physically and emotionally. Dependence on the facilities of a teaching hospital for her daughters' treatment meant that Ruby felt she had no choice but to allow her daughters to be repeatedly examined by multiple doctors. When her daughters grew they would object to the medical exams. Now an adult, one of the daughters is angry at Ruby, believing she was medically raped. For many years the relationship between Ruby and her younger daughter was very strained. Ruby is hurt by her daughter's anger and constantly reminded of the choices she made.[88]

For all of the above reasons, parents need to be supported. They are vulnerable. Pastoral responses to intersex should include provision for parents as well as intersex individuals. Both have difficulties in coping with the experience of difference. Both experience issues to do with identity and shame, although the focus of both is the intersex person.

Having explored the issues which impact on intersex persons we now turn to some issues that relate to a social response to intersex. The first of these is the agenda to get rid of gender altogether. The other is the distinction between intersex, homosexuality, and transgender, something that many people find confusing.

The Gender Agenda

Intersex calls into question the assumptions we commonly make about the binary nature of sex and gender. Part of the wider purpose of this book is to consider this issue. However, there are some who want to go beyond questioning assumptions and move towards abolishing gender altogether. I call this "the gender agenda." The advocates of this agenda use intersex and intersex persons as pawns to achieve their ends. There may be intersex persons who participate in this agenda, but many do not desire the abolition of gender. The now defunct Intersex Society of North America (ISNA) is a case in point.[89] The radical nature of this agenda is expounded here.

87. Callahan, *Between XX and XY*, 78–80.

88. Feder, *Making Sense of Intersex*, 49–53.

89. ISNA, "Frequently Asked Questions."

In her 1993 article "The Five Sexes: Why Male and Female Are Not Enough," Anne Fausto-Sterling proposed that instead of the typical Western ideal of two sexes, we should accept that there are at least five sexes: male and female, plus "herms" (true hermaphrodites), "merms" (male pseudo-hermaphrodites), and "ferms" (female pseudo-hermaphrodites). The five sexes are in fact not enough. Rather, we must conceive of a world in which there are many sexes. In this world all divisions based on sex, gender, and sexual orientation would be gone. To achieve this utopian vision requires that intersex persons openly admit to their sexual condition so that society can learn to accept different sexualities.[90] The idea is extended in a later article. Intersex persons are not the midpoint of a sexual continuum, since sex is better construed as having many dimensions. We must move beyond acceptance of gender variation to recognition that there are more variations than has previously been conceivable based on genitals alone. Reaching the goal might involve endorsing the rights of people to define their own gender or to marry whomever they please.[91]

Virginia Ramey Mollenkott goes further in her agenda than Fausto-Sterling. Mollenkott is not even satisfied with radical feminism, since feminism still operates within a sexual binary, which is too constraining. She claims that very few people are able to fit within the binary system of gender, and attempting to fit has caused mental and relational harm to many, including herself. Some have even suicided or been killed as a result of these constraints. In reality the gender binary simply does not work for intersex persons and transsexuals, so it will soon perish. Since "gender injustice" is rampant, there is dire need of a new paradigm. That paradigm is what Mollenkott calls "omnigender."[92]

Mollenkott uses the existence of intersex as proof that the idea of genitals determining social roles is a construct. This construct is unnecessary and unjust. She proposes a different paradigm, one in which people determine their own way of enacting gender, according to their own preferences at each particular moment. For Mollenkott, both sex and gender should be viewed as a continuum. Mollenkott claims that transgender persons—by which she means transsexuals, homosexuals, cross-dressers, drag queens

90. Fausto-Sterling, "The Five Sexes."
91. Fausto-Sterling, "The Five Sexes, Revisited."
92. Mollenkott, *Omnigender*, xiii–xiv, 2–5.

and drag kings, and all who identify as queer—should be embraced by society, because of their valuable contribution to understanding gender.[93]

The goal of an omnigendered society is to erase gender boundaries and to produce a society that embraces gender fluidity and diversity, making biological sex irrelevant. There is no dividing line between male and female. Each person could be unique in his or her expression of gender and, since male and female would no longer exist, the terms heterosexual or homosexual would have no meaning. Children could choose their own gender. The gendered aspects of society would be done away with: no more male and female toilets, no more segregation of children as boys and girls, no more titles such as Mr. and Mrs., no more male and female prisons, and no segregation in sporting competitions.[94]

Others also advocate a radical gender agenda. Gerald Callahan insists that intersex persons show us that there exists a continuum of possibilities lying between male and female. For Callahan intersex is not something disturbing, but something that reveals what it means to be human.[95] David Ozar contends that there needs to be room for intersex persons, and indeed every person in society, to hold a provisional gender. If society would embrace this idea as a whole, then any young person might assume a provisional gender identity until he or she had time to consider which gender most fits their personal identity.[96] David Hester takes the idea of five sexes well beyond where Fausto-Sterling left off, claiming that there are hundreds of sexes. Since for intersex persons no obvious sex exists, the sex must be constructed on the body. Gender is thereby deconstructed by the existence of intersex persons. The concept of heterosexual and homosexual desire also becomes meaningless.[97]

It is evident from these examples that there is a radical agenda to get rid of the idea that humans are made male and female. Evangelical Christians cannot condone this move. The existence of intersex people is not an indication that sex and gender are a continuum. This point will be discussed in depth in chapters 2 and 3. For now it must be stressed that those proposing this agenda are using intersex people as pawns for their political

93. Ibid., 144–80.

94. Ibid., 183–88.

95. Callahan, *Between XX and XY*, 8, 77–78, 163.

96. Ozar, "Towards a More Inclusive Conception of Gender-Diversity for Intersex Advocacy and Ethics," 30.

97. Hester, "Intersexes and the End of Gender," 218–21.

ends. Intersex persons should not be required to change the social order nor do they exist as object lessons for radical agendas. Using people this way undermines the humanity of intersex people.

Homosexuality, Transgender, and Intersex

The second matter that is important to a Christian understanding of intersex is the difference between intersex, homosexuality, and transgender. Although some intersex activists have aligned themselves with LGBTQ groups to achieve mutual political aims, this does not mean that intersex should be confused with either homosexuality or transgender. Unfortunately, in the past, intersex and homosexuality have been seen as two ways of disrupting good sexual moral order and thus in some sense mistakenly assumed to be the same thing. In the present a real confusion exists in the popular mind regarding homosexuality, transgender, and intersex. Therefore, a few words of clarification are in order.

In the late nineteenth and early twentieth centuries, the medical profession perceived intersex and homosexuality as almost the same phenomenon. Since intersex persons had ambiguous genitalia, would sexual intercourse thereby be homosexual sex? Homosexuals had perverse sex, so perhaps they had some kind of mental intersex condition or ambiguous mentality. This association meant that intersex was thought equivalent to depravity. Both homosexuals and intersex persons, then, were seen as people whose existence spelled the end of family life, because they seemed to obscure distinctions between male and female social roles.[98] However, it is unreasonable to speak about homosexuality and intersex as if they were somehow the same, one mental and the other physical. Homosexuality is not a physical fact but a sexual orientation.

To the twenty-first–century mind this distinction between homosexuality and intersex is not such a difficult concept. Perhaps more difficult to grasp is the difference between intersex and transgender. The distinction is one of biology versus psychology. Intersex individuals are intersex through biological circumstance. It is clear that no choice was made on their part to have ambiguous sexual biology. Transgender individuals consider themselves to have a mismatch between body and mind, but this is much less a biological fact and far more a psychological state. Transgender may not

---

98. Holmes, *Intersex*, 35; Reis, *Bodies in Doubt*, 56–57, 61.

be a *conscious* choice, but it does involve a desire for a gender different to biological sex.[99]

Intersex, homosexuality, and transgender are three separate matters. Intersex is a biological condition, ambiguity of biological sex. Homosexuality is a sexual orientation, sexual attraction to people of the same biological sex. Transgender individuals are those whose gender identity, their understanding of their own selves, is incongruous with their biological sex. These three should not be confused. Most conservative evangelicals are opposed to homosexuality and transgender, since these are seen as sinful. Intersex, on the other hand, is a purely biological phenomenon and should *not* be construed as sinful. Some intersex people are homosexual and some are transgender, but their homosexuality or transgenderism are separate matters to their sexual biology.

Conclusion

Intersex births happen regularly. This is a fact, whether we are aware of it or not. Since the fact is not at issue, what remains is how to respond to the reality of intersex. Several have argued that the dominant response to intersex, namely surgery, hormone therapy, and subsequent secrecy, is motivated by a desire to maintain the social order with its assumptions regarding humanity as male and female.[100] However, it is possible to respond in alternative ways. The lens through which we view intersex will ultimately determine how we respond, not just to a phenomenon, but also to people affected by intersex—intersex individuals, their parents, their siblings, their spouses, or their children. Some of those responses have been discussed in this chapter. The remainder of the book is committed to responding in an evangelical way, according to the gospel.

99. Yarhouse, *Understanding Gender Dysphoria*, 61–63.

100. Dreger, "Intersex and Human Rights," 78; Holmes, *Intersex*, 13; Preves, *Intersex and Identity*, 11–12, 20.

2

Creation and Intersex

Introduction

To be human is to be an embodied creature. Bodies are therefore an indispensible part of who we are as humans. Intersex bodies are unfortunately sometimes the site of shame. But the Bible provides good news about human embodiment, and therefore good news about the bodies of intersex persons. This chapter explores the goodness of intersex embodiment, something that derives from the goodness of God, which can never be removed or changed. Because bodies, including intersex bodies, are created by a loving God and declared to be good, there is no need for shame.

The Genesis creation account tells us about what it is to be human as well as why things are not as they should be in the world. Being a conservative Christian, I accept the biblical accounts of creation as a description, poetic as it may be, of real events and real people. Understanding the creation of the first humans is vital in order for us to understand our own selves. However, beginning with Genesis does not imply that the creation account can be understood on its own. As with the entire Old Testament, creation must be understood in the light of Jesus Christ. Creation was made for Jesus (Col 1:16), so we can only truly understand creation through him.

The God of love created humans in his image, giving humanity a unique place with respect to God and the world. Humans are made to be sexual beings. The first humans were made as male and female, two sexes sharing a common humanity. Both human embodiment and human sexuality were declared to be good. Unfortunately, the disobedience of the first humans produced fundamental changes to the world. Among those changes is the distortion of human sexuality, one expression of which is the existence of intersex variations. However, having an intersex variation cannot undo the goodness of the body, the love of God, and the value of the person. Not only is intersex embodiment still good, but God has used unusually sexed persons for his redemptive purposes.

The Love of God and Human Embodiment

God created the world out of love, because he wanted to share his eternal love with us. Nothing contrary to that love will survive his redemptive actions. Humans are made in God's image and that image is physical in form. Persons in intersex bodies are also the image of God and therefore have great dignity and value. Those bodies are good, because what God has made is good and that goodness cannot be removed. Therefore, however the body is sexed—whether male, female, or intersex—that body is good.

In the Beginning God Created the Heavens and the Earth (Gen 1:1)

God did not create the world because he was lonely. In fact he exists as a communion of three persons. The first verse of the Bible hints at this. The Hebrew word for God in Genesis 1:1 is *ĕlohim*. The Hebrew word is plural, but used with a singular verb *bārā* (he created). The Christian tradition has taken this as a veiled reference to the Trinity. The New Testament makes it explicit that there are three persons within one God: Father, Son, and Holy Spirit. Since God always existed as a communion of three persons in one God, there could be no loneliness to overcome. Therefore, we know that God had another reason to create humanity. That reason is love.

Because God was not forced to create the world, all created things, especially human beings, exist due to the gracious action of God. The creation of the universe was motivated solely by the love of God. Only because of love did God the Father desire to create creatures other than himself, and to give them continued existence. God's fatherly love is expressed through making the world to meet the needs of humanity. His love for creation is motivated by his love for the Son, mediated through the Son, and outworked by the Holy Spirit. The Spirit gives physical life and enables creatures to share in the life of God.[1] Humans are therefore creatures made purposefully by a loving God, whose desire is to include them in his love and to give them his own life.

Christian doctrine has maintained that the sovereign God created the world out of nothing. There was no preexistent matter or other power that contributed to the creation. Nothing was made that does not reflect the loving character of the triune God. Everything that God made was made with loving purposes in mind. No evil was built into the fabric of the world by some outside force. Any evil or disorder that now exists in the world cannot continue eternally, since the Creator God is sovereign over his creation. Evil, chaos, and dysfunction are not forces outside his control, but are destined to be undone by his love.

In contrast to the myths of the Ancient Near East, according to which humans were created to feed the gods and to do the work which the gods found burdensome,[2] humans—intersex and unambiguously sexed—were created because of the love of God. God created us all, however we are sexed, to be loved and cherished by him. The loving purposes of God cannot ultimately be undone by any actions of humanity, or by any evil that is present in the world. The love of God is the fundamental ground of our existence as human beings. This fact should inform all human interactions. Thus Christians must remember to act in love towards the unusually sexed, and those who are intersex can be assured that God's love is the reason for their existence.

And God Said, "Let Us Make Man in Our Image, in Our Likeness" (Gen 1:26)

Human beings are the most special creatures. They are far more important than the animal creation. Although humans and animals share many things, humans have a different relationship to the Creator. God simply commanded the animals into existence, but when it came to creating humans he deliberated and conferred with himself.

> Then God said, "Let us make man in our image, after our likeness. And let them have dominion over the fish of the sea and over the birds of the heavens and over the livestock and over all the earth

---

1. Pannenberg, *Systematic Theology Volume 2*, 20–22, 32.
2. Houston, *I Believe in the Creator*, 66.

and over every creeping thing that creeps on the earth." So God created man in his own image, in the image of God he created him; male and female he created them. (Gen 1:26–27)

Humanity is the crown and pinnacle of creation, given pride of place in the heart of God, and made to experience fellowship with God.

The Bible does not explain the content of the image of God in Genesis. However, in the history of Christian theology there have been two basic understandings of the image: either the image is found in some intrinsic characteristic of humanity, such as the soul or the intellect; or it is found in some activity that humans are to carry out, such as having dominion over the earth. The former is unsatisfactory since it reduces the human to only one characteristic and downgrades the body to an unimportant part of human being. The latter is at least based on the biblical text (Gen 1:28) but is still too narrow to be the image. Dominion over the earth is more likely a consequence of being in the image rather than the image itself.

A better way of understanding the image of God is as relationship, both vertical and horizontal. The human being in its entirety is a creature made to reflect the Creator in all of its existence and activity. To be in the image of God is to be a free creature like God, who is completely free. Freedom is, however, is not a human characteristic as such; freedom is a relationship. In order to be a free person the human must live for the other. God is free, but his desire is to give himself to us, and this he does in Christ. Therefore, there is nothing in humans as such that is like the being of God. Rather, the image of God is a relationship. Human beings exist within relationships: a relationship as the object of grace from God, and a relationship of giving to other human beings. This is what the image of God is.[3]

The human body is not excluded from the image of God, but rather humans have physical bodies. It is as embodied creatures that they are the image of God. In our bodies we are connected to the earth from which we are made and connected to other embodied humans. In this way all humanity is interdependent.[4] The Bible prohibits making idols. Some believe this is because human beings are the only image of God. So the only image of God that can be seen on the earth is the embodied human being. Without our bodies, then, we could not be the image of God. This must be emphasized given that the history of Christian theology has been dominated by

3. Bonhoeffer, *Creation and Fall*, 60–64.
4. Ibid., 79.

dualistic philosophical ideas that ignore the importance of the body and see the image of God purely in the soul.

All humans, however sexed, are created in the image of God. As such, great importance is attached to humans within the created world. Being the image of God provides humans with an unbreakable dignity. Humans are relational beings; we relate to God and to other humans. This need for relationships applies to persons who are intersex just as it does to those who are unambiguously sexed. Since humans are the embodied image of God, we cannot ever legitimately dismiss or degrade the human body. I want to emphasize that intersex people are as much the image of God as unambiguously sexed persons. Since intersex people are the image of God, they must be treated is as if they have unbreakable dignity and worth. Each intersex person is intended for relationship with God and with other people. Exclusion of intersex people, because of their unique embodiment, is a denial of the fact that they are the image of God. We must remember how valuable intersex people are to the God who made them.

And God Saw All That He Had Made
and God Saw That It Was Good

The Bible is clear that human beings are intended to be embodied creatures. In both creation accounts (Gen 1 and 2) the biblical writer makes it plain that human physicality is a good thing. Six times in Genesis 1 the writer observes that God saw what he had made and pronounced that it was good (1:4, 10, 12, 18, 21, 25). But when humanity had been created and the creation completed, God saw that it was *very* good (1:31). Humans are blessed by God. God is good and his work of creating humans is good.[5] Our physical creatureliness is good.

The second creation account tells us that humanity is made from the dust of the earth and consequently we are inescapably, constitutionally connected with the earth. We are not creatures with a soul and spirit as our core, but rather creatures who are body and soul, soul and body. The human was fashioned by the hand of God himself (Gen 2:7). This loving construction of the first human body speaks of the gentle fatherliness of the Creator. God's Spirit breathed into the human to bring it to life. This form of creatureliness makes the human distinct from the other living creatures,

5. Ibid., 68.

even while it is in many ways like them.[6] Human bodies are therefore physical and dependent on the Creator.

The goodness of creation reflects the goodness of the Creator. God is good and he created the world as an act of his goodness. He desired that his creation share in his glory and goodness, goodness that is shared between Father, Son, and Spirit eternally. The New Testament explains that the world was created in, though, and for Jesus (Col 1:16). In other words, it was made through the one who became incarnate. Because of the incarnation, Jesus is both creature and Creator, so God remains intimately connected to the physical creation.[7] The incarnation again affirms the goodness of human bodies, because the Son of God has a human body. Human bodies are therefore *very good*.

Bodies are part of the good creation of God. Bodies are good. That goodness can no more be undone than the goodness of the Creator can be undone. Whatever way our bodies manifest, whether tall or short, thin or large, blonde or ginger, having a body is good. God's creation of bodies is done with love and care. Since God is good he does not create anything other than good bodies. Even the Son of God exists in a human body eternally. However a person is sexed—female, male, or intersex—the human body is very good.

The Fall and Beyond

The world was created because of the love of God, human beings were made in the image of God, and human bodies were declared very good by God. But the reality of the world often stands in contradiction to these facts. This is a result of the fall. The Bible gives no other explanation for the brokenness of the world and the fragility of human bodies. Sin is the problem with the world, but this is the fault of us all, not the problem of particular individuals. But in another sense, the physical evil in the world does not make sense in a world that is made by a good God. Evil does not belong.

However, even though the world is broken by sin, the goodness given it by God cannot be removed. Humans are still created by a loving God; they are fearfully and wonderfully made. All humans are still made in the image of God. Although the world appears not good and life is frequently beset with difficulties and suffering, Jesus has conquered all evil and now

6. Ibid., 76, 78.

7. Gunton, *The Triune Creator*, 10; Steenberg, *Irenaeus on Creation*, 22.

enjoys the joy of God's unlimited goodness. Those who follow Jesus will also assuredly experience this when raised from the dead.

The Fall, Death, and Physical Evil

The first humans lived in a world that was in every respect manifestly good. Nothing existed that would cause humans to question the goodness of God or his world. There was nothing ambiguous or morally gray. All was good and clearly so. Yet the world in which we live is not this clear cut. It is not simply that human beings are in many ways morally corrupt, if not downright evil. The world itself is not always benevolent. Sharks are majestic and yet they attack people out surfing; the sunshine makes summer enjoyable but at the same time causes skin cancer; human bodies are wonderful in their complexity, but dreadful in their capacity to become sick and diseased. The ambiguity of the world requires an explanation.

The only explanation that the Bible provides for moral evil and physical evil is the fall of humanity into sin, that is, human rebellion against the good Creator.[8] Genesis 2:16–17 recounts:

> And the LORD God commanded the man, saying, "You may surely eat of every tree of the garden, but of the tree of the knowledge of good and evil you shall not eat, for in the day that you eat of it you shall surely die."

But in Genesis 3 Eve, and Adam with her, did eat of that tree.

> So when the woman saw that the tree was good for food, and that it was a delight to the eyes, and that the tree was to be desired to make one wise, she took of its fruit and ate, and she also gave some to her husband who was with her, and he ate. (Gen 3:6)

In this way, they plunged the whole of future humanity into a world in which both good and evil are present. No longer is the world only good all the time.

There are various kinds of evil in the world. Moral evil—unloving behavior and injustice—is something clearly connected to human sin. But physical evil often seems more remote from human actions. Humans get sick, people get cancer, horrific accidents happen, and infants are born with congenital defects. It is not difficult to connect some human bodily

8. Gunton, *The Triune Creator*, 171.

troubles to human actions. We readily accept, for example, that smoking gives people cancer. However, it is much more difficult to connect human actions to congenital defects. So how might we explain the phenomenon of ambiguous sexual biology?

The most immediate result of the fall is death. Our mortality is also the most heinous aspect of the fallen world. Death is more than physical—it affects human relationships, human work, and human reproduction (Gen 3:16–19)—but it is most certainly affects the body. Physical death is the demonstration that human bodies are corrupted by sin and no longer function in the way they were intended. This manifests in many different ways: sickness, disease, mental illness, fatigue, aging, disability, and the final dissolution of the body. Of these, possibly the most shocking and disturbing is the presence of congenital defects in newborns. Parents of infants who are born with obvious intersex conditions experience confusion, fear, uncertainty, and shock. We know in our core that congenital defects are not right; this should not happen. And yet it does.

Newborns are not responsible for congenital defects. It is inconceivable that newborns have sinned (see John 9:1–3). But let me suggest one explanation for congenital defects. Humans are utterly interconnected and affect one another in profound ways. Our interconnectedness is intended by God so that each human would benefit others and benefit from others. But this can also have negative effects. When the first humans sinned, they did not merely bring about the curse of God upon themselves as individuals, but brought it down on all humanity. Not only did Adam and Eve sever their own communion with God, and sever that communion for all their descendents, but that first sin also destroyed the bonds between one human and every other. Sin, then, is not just individual but corporate. Each person experiences the consequences of the sin of the human race. In some way this explains inherited sin and congenital defects, because we all experience the consequences of humanity's sin, even before we are born, due to our intrinsic connection to other humans.

These arguments provide some kind of explanation and yet, no matter what careful explanations are offered, in the end it is clear that there is no rational reason for evil to exist in God's good world. It does not and cannot make sense. Evil has come into God's good and orderly world and it is opposed to everything that God has created to be good and to experience only good. But although we cannot understand why God has permitted suffering and evil, yet we can be confident that he will overcome evil and undo suffering for those whose trust is in him.[9] The same chapter of Genesis that describes the fall also offers hope that the situation will be undone (Gen 3:15). Redemption has come in the person of Christ.

Intersex in humans is generally considered a congenital defect of sexual anatomy. Although not all intersex people are distressed by their bodies, some yearn for an explanation. British Christian intersex man Matthew asks, "I thought, if God was so loving, why did he make me like this?"[10] God does indeed love intersex people and they are created by him. I can offer no explanation for congenital defects of bodies and sex organs other than the fact that human sin has changed the nature of the world and the nature of human bodies. It is not the individual sin of intersex people that resulted in their unusual biology; it is the result of every person's sin. We all contribute to the physical evil in the world; every sinner is responsible for intersex births.

Humanity beyond the Fall

Given that the fall has radically changed humanity and the world, what can we say about humanity in the post-fall world? First and foremost, God is good and he has not changed. The world is good because it was created by a good God. It is still dependent for its existence on the good Creator. Since God the Creator is continually Lord over his creation, the created world cannot be called other than good.[11] This fact has implications for humanity in the fallen world.

First, individuals are still a direct creation of God. When the psalmist speaks of his own being, he believes that he is a creature made by God.

> For you formed my inward parts; you knitted me together in my mother's womb. I praise you, for I am fearfully and wonderfully made. Wonderful are your works; my soul knows it very well. My frame was not hidden from you, when I was being made in secret, intricately woven in the depths of the earth. Your eyes saw my unformed substance; in your book were written, every one of them, the days that were formed for me, when as yet there was none of them. (Ps 139:13–16)

9. Pannenberg, *Systematic Theology Volume 2*, 16.

10. Cornwall, "British Intersex Christians' Accounts," 226.

11. Bonhoeffer, *Creation and Fall*, 45.

For the psalmist, it is not a contradiction that he was conceived and grown within his mother's womb and yet still made by the Creator. God knew him before he was born and formed his body as it grew in the womb, hidden from sight.[12] This fact is not undermined by the reality that humanity is born into a world under sin. The psalmist was most certainly aware that humans inherit sin from the womb (Ps 51:5), but this can in no way undo the wonder of being created by God.

But we know that even as humans are fearfully and wonderfully made by God, children are born with birth defects or even sometimes stillborn. Why does God create in this way? God accepts human beings as sinners and fallen. He does not undo the fallenness of humans or undo the consequences of human actions, but rather acts within the broken world that the first humans brought about. He does not break the laws that now operate in the fallen world, but works within them.[13] The mechanisms that guide conception, gestation, and birth are now corrupted by the sin of the human race as a whole. Therefore, some people have birth defects and some die in the womb, because of the brokenness of the world, while at the same time they are wonderfully created by a loving God.

What about the image of God? The Old Testament mentions the image only twice after Genesis 1:26–27 (Gen 5:1–31; 9:6), but the New Testament still assumes that sinful humans are the image of God (1 Cor 11:7; Jas 3:9).[14] However, sin has impact on the human person in significant ways. This is why Reformed theologians in particular say that the image of God is destroyed by the fall and only restored again in Christ. It is through the gospel that humanity is again changed into the image of God (Eph 4:23; Col 3:10). When the Reformers claimed that the image is destroyed by the fall, they were concerned to emphasize that sin has entirely destroyed original human righteousness. God is no longer able to see his reflection in the human, because sin has marred that reflection beyond recognition.

The continuance of the image and the claim that it is destroyed by the fall seem to be contrary claims. Yet the two make sense together when we consider Jesus. Jesus Christ is the primary image of God (1 Cor 4:4; Col 1:15). The image of God was not fully made known until Jesus came and showed us what God is like. Since humans are created in Christ (Eph 2:10; Col 1:16), Jesus the true image of God upholds the image in other humans.

---

12. Brunner, *Man in Revolt*, 89.

13. Bonhoeffer, *Creation and Fall*, 139.

14. Mathews, *Genesis 1–11:26*, 170–71.

God has chosen humans as *the* creatures among all his creatures to be in a special relationship with him. Since God is faithful, the image cannot be erased.[15] Therefore, human dignity cannot be expunged.

The world beyond the fall is still good, but not everything appears to be good on this side of the fall. Indeed, in some events and some circumstances it is difficult to see the good. Being a Christian does not exempt a person from disaster and difficulty. God's goodness is sometimes only "seen" in the invisible realm where "we know that for those who love God all things work together for good, for those who are called according to his purpose" (Rom 8:28). It is only because Jesus has dealt fully—in his cross and resurrection—with all that threatens the goodness of creation, that we can understand in truth that creation is utterly good.[16]

Jesus lived in the fallen world, experienced things that do not seem to be good, and died upon the cross. There Jesus experienced all the evil and brokenness of the fallen world in its most intense form. But he has risen from the dead and is now seated at the right hand of God. From the perspective of Jesus, it is not that one day God will make all things good and explain all the mysteries of his providential work in the lives of his people, but rather Jesus has arrived at the place to which all providence is headed. From there he experiences all the goodness of God in everlasting peace and joy (Acts 2:28; Heb 12:2). Therefore, God is not merely one who understands our experience of pain and suffering, but he has conquered it fully.[17] For this reason, we can have confidence in the goodness of creation.

Humans beyond the fall are still creatures of God, created by the loving Creator. All people, whether intersex or unambiguously sexed, are fearfully and wonderfully made. This fact has not been changed by the fall. All people, regardless of their sexual biology, are made in the image of God. That image has been distorted by sin, but not destroyed, because it is upheld in the person of Christ. Since the image of God is dependent on Jesus and not on our sexual biology, being intersex cannot diminish a person's worth. Although for many people who are intersex the world does not always appear to be good, each can be confident in the faithfulness of God through Christ. From the perspective of Jesus the goodness of the world is never in doubt.

---

15. Gunton, *The Triune Creator*, 206–8.

16. Ibid., 157.

17. Forsyth, *The Cruciality of the Cross*, 61.

Being Male and Female

So far in my examination of human origins, I have emphasized that humans are created in love by a good God. Humans are created in the image of God, endowing them with dignity and honor. Human embodiment is good and human bodies should never be dismissed as unimportant. But up to this point, I have not considered what appears to be a major issue with regard to intersex persons, the creation of humans as male and female. This is, so to speak, the elephant in the room. The Bible is unambiguous about the creation of male and female in the beginning. Some have even suggested that male and female is the image of God. There is most definitely a theological challenge here. However, any theological conclusion must be drawn from the grace of Christ Jesus.

Humanity as Male and Female

The first chapter of Genesis situates humans within the larger creation of the cosmos and the plant and animal life. Chapter 2 allows us to see humanity up close and personal, at the center of the creation. In both of these stories it is plain that humans were originally made as male and female.

> Then God said, "Let us make man in our image, after our likeness. And let them have dominion over the fish of the sea and over the birds of the heavens and over the livestock and over all the earth and over every creeping thing that creeps on the earth." So God created man in his own image, in the image of God he created him; *male and female he created them.* And God blessed them. And God said to them, "Be fruitful and multiply and fill the earth and subdue it, and have dominion over the fish of the sea and over the birds of the heavens and over every living thing that moves on the earth." (Gen 1:26–28)

In chapter 1 the creation of humanity follows the creation of land animals. Yet humans are not identical with animals. Animals are made "according to their kinds" (1:24), but the unique designation for humanity is that they are male and female. In this sense, only humans are created with sexuality. Significantly, human sexuality seems in some way connected to being created in the image of God. The two lines of poetry in Genesis 1:27 are an inverted parallel:

In the image of God *created he him*;
Male and female *created he them.*
The shift from singular "him" to plural "them" underscores the two sexes
and denies the creation of an androgynous human. Humanity was created
as two different and complementary sexes, not one human with two sexes
in one body. Right from the beginning humans were differentiated into
sexes and yet unified in their humanness.[18]

Human sexuality was included within the pronouncement that all that
was made was good. Humanity was given the divine blessing for procreation and made male and female so that this would be possible.[19] Yet to
think of humans as male and female for the purpose of reproduction alone
reduces humanity to one function. As Paul K. Jewett points out:

> The procreative function of the sexes, important as this may be,
> is only one among many aspects of a complex, creative, dynamic,
> post-fall human fellowship, a fellowship which expresses itself in
> and through a variety of specific relationships to the benefit of
> both the individual and society as a whole.[20]

In chapter 2 the point of view changes; the focus narrows to the creation of humanity. "Then the LORD God formed the man of dust from the
ground and breathed into his nostrils the breath of life, and the man became
a living creature" (Gen 2:7). For the man, the LORD God planted a garden,
full of beauty and plentiful food, well watered and rich with precious metal.
The man was given a meaningful task to do and freedom to eat whatever
fruit he desired in the garden (except one). The man had everything possible to make him happy in the brand new world. Yet this was not enough.
The man was alone. "Then the LORD God said, *It is not good* that the man
should be alone; I will make him a helper fit for him'" (Gen 2:18).

Why would the man who was with God and in his care need a partner,
a helper? Animals are physically like humans and yet none of them proved
to be a suitable helper. Instead, God put Adam to sleep and created a suitable helper for him from his side. Since the woman was formed using a part
of Adam, he knew that he was bound to her in a profound way. "Then the
man said, 'This at last is bone of my bones and flesh of my flesh; she shall
be called Woman, because she was taken out of Man'" (Gen 2:23). They
belonged to one another; they were in some sense one, while still being two

18. Trible, *God and the Rhetoric of Sexuality*, 15–18.

19. Loader, *Making Sense of Sex*, 9–10.

20. Jewett, *Man as Male and Female*, 49.

persons. The man and the woman would come together again to become one (2:24), and yet this oneness does not eradicate or even diminish their individuality. Rather, their oneness was grounded in their difference from one another.[21]

The biblical picture seems clear. Humans were created in two sexes. There are, however, two opposing arguments about the creation of male and female. Sally Gross argues that Adam was a hermaphrodite. Genesis 1:27 makes a shift from God creating the human, referred to in the singular as "him," to creating male and female, referred to in the plural as "them." Since this is an unusual use of singular and plural, some rabbis have tried to explain it. In the *Bereshit Rabba* we find: "Rabbi Yirmiyah [Jeremiah] ben 'El'azar said: When the Holy One Blessed be He created the primal man ['the primal Adam'], he created him an androgyne, and it is therefore said: 'male and female he created them'" (Gen 1:27).[22] Gross acknowledges the anecdotal nature of the comment. However, because the rabbis were so concerned for accuracy in the details, the very existence of such a comment in this context suggests that we cannot use Genesis 1:27 to make a clear-cut division between male and female.[23] Her conclusion is that "Hermaphroditism should perhaps be seen as a reminder of the 'original innocence' and perfection before sin distorted it."[24]

The second argument is from an Eastern Orthodox tradition. Many early Eastern church fathers saw sexuality as a result of the fall, believing that sin is principally lust. Humans are male and female because of the fall, rather than because of created intent. According to Modern Russian theologian Nicholas Berdyaev, humans are both masculine and feminine within one person, but the proportions of each differ. Men are not human without some feminine aspects, nor are women human who have no masculine aspects to their personality. The feminine is "communal and cosmic" and the masculine "personal and anthological." Since the fall, the two principles seek to unite and also oppose one another. Consequently humans are unable to be at peace with one another while they remain sexual beings.[25]

Neither argument matches up well with the Genesis account of creation. The creation of male and female is clearly recorded as taking place

21. Bonhoeffer, *Creation and Fall*, 95–97.

22. Cited in Gross, "Intersexuality and Scripture," 71.

23. Ibid., 70–71.

24. Ibid., 74.

25. Jewett, *Man as Male and Female*, 25–26.

before the fall. Therefore, human sexuality cannot be attributed to the fall. Sin encompasses far more than sexual lust and should not be reduced to, or centered on, one aspect of human behavior. Genesis 1:27 does not imply a hermaphrodite human. The text is intended to give us the opposite impression. It does not merely say explicitly that humans were created as male and female, but the following verse implies two human sexes. Humans were commanded to "be fruitful and multiply and fill the earth" (Gen 1:28), a command that requires the procreative capacity of two sexes. There are also theological reasons why these arguments are incorrect, and these will be discussed in the next chapter.

The fact that Genesis clearly states humans were created as male and female certainly presents a problem, since there are without doubt intersex individuals in the world. There must be some explanation for this fact. We must tread carefully in this exploration of Scripture, since the bare facts of the Bible can be read as condemnation on humans who are neither clearly male nor clearly female. Yet as I have intended from the start, this chapter is not merely about the bare text of Genesis. It must be understood in the light of Christ, who has come into the world "full of grace and truth" (John 1:14). The love of God does not condemn intersex people for their biology.

Is "Male and Female" the Image of God?

There is a second issue regarding intersex that arises from the creation narrative. There appears to be some kind of connection between being created in the image of God and being created as male and female. Some theologians have stated that humanity as male and female *is* the image of God. However, the New Testament states that Jesus Christ—not male and female—is the image of God.

The most significant claim that humanity as male and female is the image of God comes from twentieth-century theologian Karl Barth. Since God is in no way solitary, but rather exists as a Trinity, solitary humanity cannot reflect God. Thus humanity must by nature be "fellow humanity." The primary way in which humans express this fellow humanity is as male and female. The significance of this is found first in marriage between male and female, since according to the Old Testament marriage represents the union between Yahweh and Israel, and according to the New Testament the union between Jesus Christ and the church. When Genesis 1:26–27 speaks of humans being in the likeness of God, it explains this by saying that humans are created male and female. Humanity is like God by existing in relationship, just as God exists in relationship, rather than being alone.[26]

There are pluses and minuses to Barth's argument. Positively, male and female as the image of God is an improvement on the historically predominant view that the image of God is found in reason. That view has for centuries been used to exclude women from the image of God, because they are "irrational." Barth's view gives women an equal share of the image.[27] Barth's thesis has other positives. Most importantly, it gives place to community over isolation and individuality. Instead of a Cartesian understanding of the human person which requires only an "I"—"I think, therefore I am"— humans must be in fellowship to be truly human. The image is the whole person, rather than only one aspect, such as the soul or reason. Sexuality holds a place of honor rather than being inexorably associated with sin.[28]

Negatively, Barth has overextended his interpretation of Genesis 1:27. Although the two descriptions of humanity—"image of God" and "male and female"—are adjacent, they do not have to be synonymous. Moreover, Barth has not simply read the text of Genesis, but rather used constructive interpretation, adding to the text. The relationship between male and female in the creation narrative is significant without doubt. However, it is not the image of God as evidenced by the fact that none of the other references to the image make a direct connection to male and female.[29] Male and female may be the primary way in which humans act together as fellow humanity, but intersex people can also act as fellow humanity. Intersex persons are capable of entering into reciprocal relationships in the same way as unambiguously sexed persons.

Further, the New Testament makes clear that the image of God is primarily and centrally found in the person of Christ. Second Corinthians 4:4 and Colossians 1:15 state this explicitly. We must therefore draw our main understanding of the image from Christ rather than from Old Testament passages or from speculation based on cultural assumptions about humanness. God's goal is that believers be transformed into the image of Christ (Rom 8:29). This process involves being changed from one degree of glory to another (2 Cor 3:18) and putting on the new self (Col 3:10). The final mention of the image of Christ is 1 Corinthians 15:49, which tells us that

26. Barth, *Church Dogmatics III.4*, 116–17.

27. DeFranza, *Sex Difference in Christian Theology*, 148.

28. Stephenson, "Directed, Ordered and Related," 448.

29. Berkouwer, *Man*, 72–73.

believers will "bear the image of the man from heaven." None of these passages suggests in any way that the image of God is male and female.

Indeed, even if Barth is correct—and this is unlikely—this does not necessarily imply that people who are unambiguously sexed are closer to the image of God than those who are intersex. The fall of humanity has impacted humans in multiple ways. If the image of God involves the whole person, then sin also affects the whole person. Humans do not image God in the clear and untainted way that Adam and Eve did when first created. Sin in any form cannot reflect the holiness of the Creator. But sin has affected human sexuality as well as humanity's relationship with the Creator and human-human relationships. Therefore, as the next section explores, the male-female relationship does not clearly reflect the dynamic of relationships within the Trinity anymore. The image of God can only be restored in Christ.

Sexuality after the Fall

Humans were made male and female in the beginning. At first glance, this fact places intersex people outside the norm for humanity. However, things are not that simple and theology must not be that simplistic. We don't live in a sin-free world anymore. We don't live in a world where everything is as it should be. Human sexuality is no longer as it should be. This fact does not apply only to the sexuality of intersex people, but also to the unambiguously sexed. Sexuality has been distorted by the fall in many different ways, some physical, some social, some emotional. This does not mean that intersex was intended by God. It does not prove that sex is a continuum. But intersex is most definitely not *the* distortion of sexuality, only one of many. The unambiguously sexed must accept that their sexuality is also affected by the fall.

The Corruption of Sexuality

Human sexuality changed significantly after the fall. After Eve and Adam disobeyed God, a number of things changed in the world. Sexuality was affected along with every other aspect of human life. Humans experienced shame about their bodies; people are no longer comfortable with their physical sexuality. The fall also produced a significant change in the male-female dynamic. Men began to dominate women, but women still need men. Instead of unity between male and female, there is now co-dependency. Sexuality is thus very much distorted by the fall and has become something that was never intended by the Creator.

The first thing which eating the forbidden fruit produced was shame.

> Then the eyes of both were opened, and they knew that they were naked. And they sewed fig leaves together and made themselves loincloths. And they heard the sound of the LORD God walking in the garden in the cool of the day, and the man and his wife hid themselves from the presence of the LORD God among the trees of the garden. But the LORD God called to the man and said to him, "Where are you?" And he said, "I heard the sound of you in the garden, and I was afraid, because I was naked, and I hid myself." (Gen 3:7–10)

Before Adam and Eve sinned they were completely comfortable with their nakedness (2:25). But since sin entered the world, humans feel shame when naked, even when they are alone. We feel shame since we no longer trust our Creator absolutely. Furthermore, Adam and Eve tried to cover themselves by putting on loincloths, indicating they were ashamed of their *sexuality*.[30] Adam and Eve became conscious of their own bodies and as a result they became "divorced," cut off from one another.[31]

God's punishment on Eve instituted a new relationship between men and women. "To the woman he said, 'I will surely multiply your pain in childbearing; in pain you shall bring forth children. Your desire shall be for your husband, and *he shall rule over you*'" (Gen 3:16). Adam's new power over his wife was exemplified in naming her as he had previously named the animals. This act effectively put Eve on a par with the animals. Naming the woman was the first act of a new dysfunctional relationship in which the man attempted to dominate the woman by defining the boundaries of her life. No longer was she able to act as a free person whose task was to rule together with the man, as helper to the man and co-regent of the world. Adam's naming of Eve narrowed her life down to reproduction. The place of the woman was reduced to nurturing as if this were her only natural sphere.[32]

---

30. Luther, *Luther's Works* 1, 167.

31. Barger, *Eve's Revenge*, 137.

32. Ibid., 138.; Luther, *Luther's Works* 1, 219.; Trible, *God and the Rhetoric of Sexuality*, 133.

The fall affected the previous unity of the first couple. Before sin the sexual difference was the basis of unity. After sin it became the basis of opposition to one another. Previously the man had gloried in the woman, who was flesh of his flesh and bone of his bone (Gen 2:23). After sin he turned against her. He tried to justify himself by blaming the Creator for making the human couple male and female. The woman's response was different. She did not blame the man, but distanced herself from him. The differences that united the two as one flesh became opposites; the sexes were no longer complementary but opposite.[33]

The woman continued to desire the one-flesh union for which the man and woman are made. But sin distorted this desire so that the woman looked to the man for what she could only receive from the Creator. Instead of getting what she desired, she was given over to domination by the man, both emotionally and physically. Women try to attain their way by manipulation, particularly in regard to sex, thus making women even more under the power of men.[34] It is, then, not surprising that we have aphorisms such as "women give sex to get love and men give love to get sex." The relationship between the sexes is not one of mutual self-giving love as in the trinitarian life, but rather co-dependence. Both men and women lose in this "game." The woman loses because she is dominated by the man. The man loses because he no longer has a "helper fit for him," but rather a subordinate who does his bidding.

What, Then, Can We Conclude about Intersex?

After exploring the Genesis accounts of creation there are some conclusions that we can make regarding intersex. These conclusions are both negative and positive, and something that those who are unambiguously sexed should consider with humility. First of all, humans were originally created to be male and female, two sexes and two corresponding genders. There is no room in the Genesis accounts of creation for positing a third sex or a third gender. We cannot, therefore, make intersex into a third sex or a third gender. Nor, according to the biblical accounts, is it legitimate to suggest that sex or gender exist on a continuum. This may seem like bad news for intersex persons. There is, however, more to be said. It is impossible to ignore the fall and its impact on human sexuality.

33. Trible, *God and the Rhetoric of Sexuality*, 116–20, 28.

34. Barger, *Eve's Revenge*, 139.

Sexuality, broadly speaking, has been distorted in many ways by the fall. The fall has clearly impacted the relationship between male and female. Although there is no discussion in Genesis of physical changes to sexuality, there are two reasons why we should include physical distortions of sex within the list of sexual changes after the fall. First, Eve was punished by God in regard to the physical act of reproduction (Gen 3:16). The second reason is the now universal presence of death (Gen 2:17). While, without doubt, death encompasses more than the physical, it most certainly involves a physical dissolution of the body. We begin to experience that physical dissolution even before birth. The combination of the two facts—creation as male and female, and physical distortions of sexuality after the fall—leads me to conclude that intersex conditions are *one* (not *the*) result of the fall.

Several people have, however, argued that intersex is actually a natural variation of sexuality. There are many examples of animal life that exhibit intersex conditions or true hermaphroditism. For instance, over a hundred species of fish are either hermaphrodites their whole lives or they change sexes at some point.[35] Humans, however, are not fish. Humans are made in the image of God and consequently have dignity that can never be attributed to fish. We eat fish, keep them as pets, and feed them to other pets. None of these things would be appropriate to do to a human being. Any attempt to justify the idea that it is "natural" for humans to be intersex based on the sexuality of fish is not valid.

It may or may not be natural—in the sense of created intent—for fish to be hermaphrodites. It is, however, definitely not the created intent that humans be hermaphrodites. It is false to declare that everything that occurs naturally is intended by the Creator to be that way. Not everything that exists necessarily ought to exist. It is simply untrue that intersex is "incontrovertible physical evidence"[36] that male and female are not the only natural ways of existing as human. Not all difference is bad, but neither is all difference the will of God.[37]

It is possible to see intersex as a particular bodily fact that occurs all over the world without assuming either that it is the created intent of God or that the person who is intersex is more or less sinful, or more or less loved by God, than a person who is unambiguously sexed. Intersex is simply *one* distortion of sexuality. No one has escaped the distortion of our

35. Callahan, *Between XX and XY*, 110.

36. This is Kessler's phrase. See Kessler, *Lessons from the Intersexed*, 31.

37. Lebacqz, "Difference or Defect?," 24.

sexual being. It is not appropriate for people who are unambiguously sexed to act as if they had achieved their own sexual biology. Paul's question to the Corinthians is relevant here: "For who sees anything different in you? What do you have that you did not receive? If then you received it, why do you boast as if you did not receive it?" (1 Cor 4:7). We cannot congratulate ourselves on something in which we had no say. On the other hand, the person who is intersex should not feel ashamed of that which they did not choose. Unusual sexual biology cannot negate the love that God has for his human creation, since that love is upheld by the Father's eternal love for the Son in the Spirit.

Hope for the Future

It is totally unsatisfactory to leave the impression that being intersex is a second-rate way of being, as if people who are intersex are less human or less valuable than those whose sexual biology is unambiguous. It is the biblical pattern that God uses people who are unusual or considered small or weak to accomplish his purposes. This is the case with people who have unusual sexuality. Sin is not the last word in the creation account in Genesis, but rather God has promised a Redeemer. The way in which that Redeemer would come to be born is through a number of women whose sexuality is outside the norm. This fact should encourage people with intersex conditions, because it opens up the possibility that God will use those with ambiguous biology for his glory.

The story of redemption begins very early on, even before God punished Adam and Eve for their disobedience.

> The LORD God said to the serpent, "Because you have done this, cursed are you above all livestock and above all beasts of the field; on your belly you shall go, and dust you shall eat all the days of your life. I will put enmity between you and the woman, and between your offspring and her offspring; he shall bruise your head, and you shall bruise his heel." (Gen 3:14–15)

The story of the Bible is the story of the work of God to bring this promise to fruition. The serpent—that the New Testament calls the devil or Satan (Rev 20:2)—opposes humanity, but ultimately God sent his Son in order to crush the head of the serpent. God promised that the Redeemer would be born of a woman. The final woman in the saga was the Virgin Mary, who gave birth to the Messiah. The process of bringing the Redeemer to birth, however, began long before this. This section considers five people who were significant in laying the foundation for the coming Savior. Each is someone whose sexuality was unconventional: Sarah, Ruth, Tamar, Rahab, and Bathsheba.

The call of Abraham is considered pivotal in the history of redemption (Gen 12:3), but Sarah is also vital, since the covenant was established with her son (Gen 17:19–21). But Sarah's sexuality was problematic since she was unable to bear children (Gen 16:1). "The way of women had ceased to be with Sarah" (Gen 18:11). In other words Sarah was both barren and past menopause. Her inability to have children meant that she could not fulfill the mandate of Genesis 1:28 to "be fruitful and multiply" and thus was not a complete woman. At least this was how Sarah perceived herself and therefore she forced her Egyptian maid Hagar onto Abraham, in order to build a family through the surrogate (Gen 16:1–2).

Sally Gross suggests that both Abraham and Sarah may have been intersex. According to the Babylonian Talmud (an ancient Jewish text):

> R. Ammi stated: Abraham and Sarah were originally of doubtful sex [*tumtum*]; for it is said, Look unto the rock whence you were hewn and to the hole of the pit whence you were digged, and this is followed by the text, Look unto Abraham your father, and unto Sarah that bore you.

> R. Nahman stated in the name of Rabbah b. Abbuha: Our mother Sarah was incapable of procreation; for it is said, And Sarai was barren; she had no child, she had not even a womb.[38]

Tumtum is an intersex term. It means that the person has no genitals and thus cannot be assigned a sex. Sarah's lack of a womb suggests that she may have had complete androgen insensitivity syndrome (CAIS).[39] Although the Babylonian Talmud does not have the authority of inspired Scripture, it may be plausible that Sarah was intersex. CAIS would explain her barrenness. On the other hand, there are other possible explanations.

Even if Sarah was not intersex, she was still incapable of having a child, something that is unusual for women. This makes Sarah a woman whose sexuality was outside the norm. Despite her sexual function being outside of the created intent as given in Genesis 1:28, Sarah is a pivotal figure in the history of redemption. She was used of God, not because she was sexually

38. "Tractate Yebamoth Folio 64a."
39. Gross, "Intersexuality and Scripture," 71–72.

ordinary and able, but because she was not what is expected by creational standards. The New Testament has nothing but positive things to say about Sarah. Between Ishmael, the son of the "normal" woman, and Isaac, the son of the "defective" woman, Sarah's is the more significant son, since he is the son of promise (Gal 4:23). Sarah herself is a woman commended for her faith. "By faith Sarah herself received power to conceive, even when she was past the age, since she considered him faithful who had promised" (Heb 11:11).

Ruth is another Old Testament saint in the line of the Messiah. Ruth's son Boaz was a direct ancestor of King David (Ruth 4:17–22) and therefore of Jesus. But, like Sarah, there is good evidence that Ruth was also barren before her marriage to Boaz. In ten years of marriage to her first husband (1:4) she produced no children. Therefore, she married her husband's nearest relative, Boaz, "to perpetuate the name of the dead in his inheritance" (4:5), according to cultural practice (Deut 25:5–6). The other reference that implies previous barrenness is Ruth 4:13—"So Boaz took Ruth, and she became his wife. And he went in to her, and the LORD gave her conception, and she bore a son." Despite Ruth's previous inability to bear a child, she was given a part in the redemption history of Israel and ultimately the whole world.

Ruth is mentioned in Matthew's gospel as part of the genealogy of Jesus (Matt 1:5). Matthew mentions three other women in the genealogy of Jesus: Tamar (1:3), Rahab (1:5), and Bathsheba, the mother of King Solomon (1:6). Unlike Sarah and Ruth, the sexuality of these women is unusual because their sexual behavior was outside socially and morally acceptable parameters.

The story of Tamar is found in Genesis 38. Tamar was the daughter-in-law of Judah. When her first husband died, the next brother refused to produce a child for his brother, and the LORD put him to death. But Judah would not allow his other son to marry Tamar, because he was afraid that the third son would also die. So Tamar disguised herself as a prostitute in order to get Judah to sleep with her, and she fell pregnant by him. When Tamar's behavior was revealed, it also became known that it was Judah who had slept with her. Then Judah said, "She is more righteous than I, since I did not give her to my son Shelah" (Gen 38:26). If Tamar had not transgressed the sexual boundaries, the line of Judah would not have led to the Messiah.

Rahab was a prostitute in Jericho (Josh 2:1). However, it is not for her sexuality that she is remembered, but for her faith. Rahab hid the spies that Joshua sent to spy out the land (2:2–7). She did so because she believed that the LORD would enable Israel to conquer the land of Canaan (2:8–11). Her faith was significant enough that she is mentioned in the New Testament twice outside the genealogy in Matthew. She is given a place in the faith chapter of Hebrews. "By faith Rahab the prostitute did not perish with those who were disobedient, because she had given a friendly welcome to the spies" (Heb 11:31). She is also mentioned by the Epistle of James (2:25) as among the justified. Her transgressive sexuality did not prevent her from becoming a woman of faith, who is held up as an example to others.

Finally, Bathsheba was the mother of Solomon. She would not have given birth to the son who would take over the throne from David, and who would be next in the royal line that led to the birth of Christ, if she had not committed adultery with King David. David's actions are condemned by the prophet Nathan, and the son conceived through that adultery died (2 Sam 12:1–23). That Solomon was later born out of this union and chosen as the next king of Israel cannot be attributed to anything other than grace.

Grace is a defining feature of God's dealings with all three women who transgressed sexual boundaries. The Bible does not condone the sexual behavior of the women who acted as prostitutes or committed adultery. Sexual intercourse belongs within the marriage relationship. However, the sexuality of these women did not prove to be a barrier too great for God. The three women whose sexual behavior was sinful are included in the genealogy of Jesus along with Ruth, whose biological sexuality was dysfunctional. Together with barren Sarah, barren Ruth was a woman used in a deeply significant way.

None of these women can be definitively demonstrated to be intersex, although potentially this is true of Sarah. Yet they were all sexually outside of the norm. Every one of them has a vital place in the history of redemption. If these women of unusual sexuality can be so significantly used of God, then there is no reason why people who are intersex cannot expect to be used of God, if they are people with faith in Jesus. The God and Father of our Lord Jesus Christ does not work only in and through those who are sexually normal or unambiguous. The fact that a person is not biologically clearly male or clearly female is no reason to assume that he or she is of no significance. The perfect people who we imagine God is waiting to use for his purposes do not exist. Given that Sarah, and the women given place in

Matthew's genealogy, were all on the sexual margins in some way, it may be that people who are intersex are more likely to be used significantly by God than those who are unambiguously sexed.

Conclusion

The Genesis account regarding the creation of humans is very important. It explains much about humans as God intended and humans as we now exist. Humans were made because of God's goodness and love to reflect God in bodily form. Humans were intended to be male and female, to be two sexes working together as one humanity. Sin has distorted human bodiliness and sexuality, but sin is never the last word. The grace of God overcomes sin. God's grace is evident through his use of the unusually sexed in his plan of redemption.

Although this chapter has explored what the creation accounts say about human sexuality, there is a great deal more that needs to be said about sex and gender. It is impossible to ignore the present cultural push to change our understanding of sexuality. Therefore, the chapter that follows takes up more questions about sex and gender and the implications for intersex people.

3

Sex, Gender, and Intersex

Introduction

THIS CHAPTER EXPLORES SEX and gender from a theological perspective. The previous chapter considered human sexuality as discussed within the creation narratives of Genesis 1–3. A theological perspective, however, reflects on the Bible as a whole and considers the views of Christians through the centuries. This is vital in the present cultural climate, since our culture is undergoing a transition to a post-Christian society. There is a cultural shift in ideas about almost every aspect of life, including sexuality. A distinct agenda exists within sections of Western culture to radically alter our understanding of gender and even to eradicate the distinctions between the sexes. The emerging awareness of intersex persons is being exploited to solidify an unbiblical understanding of human sexuality. Since human sexuality is such a fundamental aspect of our humanness, this chapter considers it in more detail than the previous chapter, by touching on many significant theological ideas.

The chapter is divided into four parts. The first part considers the arguments for a gender continuum. Some insist that the existence of gender is the cause of much injustice in the world. Gender injustices certainly exist, but the abolition of gender is not the right response to the problem. Christians need to be aware of the pagan roots of this cultural agenda. The second part is devoted to a theological understanding of sex and gender. This explains that two human sexes parallel the relationships within the

Trinity. Two human sexes also exist to picture of the ultimate destiny of humanity, the wedding supper of the Lamb.

The second half of the chapter returns to a consideration of intersex, specifically the positive place of intersex people within the kingdom of God. Part three considers salvation, gender roles, and discipleship. These are not different for intersex people than for the unambiguously sexed. The final part discusses how the intersex person fits within the world in which male and female predominate. The commendation of the eunuch, the value of singleness, and the centrality of relationships are emphasized here. Genitals are not what make us human persons.

One final matter needs to be clear at the outset. Although in this chapter I am upholding the biblical ideal of humanity as male and female, it is important to note what I am *not* endorsing. Much of the literature opposed to a sexual binary is concerned about the implications of this ideal for intersex infants. Many intersex infants have been subjected to genital surgery in order to uphold that ideal.[1] In insisting on the significance of two biological sexes and two corresponding genders I am not thereby implying that genital surgery on intersex infants must be the result. If surgery is required or desired, this is not generally an emergency. I am not endorsing genital surgery on infants to make others feel comfortable. My concerns lie elsewhere. Additionally, I am not interested in trying to make intersex persons invisible or to insist on secrecy. From the outset I have stated that I want to defend the value and significance of ambiguously sexed people. The church must show the love of God to every human being, regardless of class, race, sex, or gender. Inclusion, not ignorance or pretense, is my aim.

Secular Trends, Gender Injustices, and Responses

Many people have found flaws in the binary-sex system on which our society is based. Some of those criticisms have a degree of validity. There are real injustices based on gender. Women are treated as inferior to men. Men have unrealistic expectations imposed on them. This, however, does not imply that we should adopt a system of gender fluidity or completely abolish gender distinctions. Gender fluidity is not as positive as it is claimed to be. Indeed, the origins of the arguments for gender fluidity and the abolition of

---

1. For example, see Chase, "Surgical Progress Is Not the Answer to Intersexuality," 147; Mollenkott, *Omnigender*, 1–2.

gender are not Christian, but pagan. Evangelical Christians cannot rightly deny the significance of either biological sex or gender expression.

The Problem with the Sex/Gender Binary?

Virginia Mollenkott devotes an entire chapter to gender injustices. She perceives the problem to be what she calls the "binary gender construct." There are many evident injustices perpetrated against men, women, and children. Women are expected to be passive, reliant, and nurturing; men are required to be active, autonomous, and strong. Women are mainly confined to the private sphere and men to the public sphere. Women are pressured to gain the status of marriage, and men are derided for anything resembling a feminine characteristic. Men are privileged in business, media, politics, and law. Most boys are unable to think of any advantages to being a girl, even though girls have no trouble thinking of the advantages of being a boy. Women are underpaid and vilified for having careers or ambition. Men define what is significant in the lives of women; women allow themselves to be co-dependent. Men also lose out because they are socialized to ignore their own pain and to suppress their tenderness and compassion. Physical strength and violence are expected for males and many are killed as a result.[2]

What are we to make of these issues as Christians? It is difficult to deny that injustices exist. Men and women do not act in loving ways towards one another. This is an unfortunate consequence of the fall. Cultural expectations on men and women are often constraining; women have been denied the use of their gifts as human beings and men are expected to act in ways that quash all sensitivity. The heterosexuality and complementarity trumpeted by conservative Christians does not always produce loving marriages and happy families. The question is not whether these problems exist, but how to respond to them. Mollenkott's solution is to eliminate the "binary gender construct" altogether. Mollenkott is hardly alone in this desire. New guidelines in Canadian schools state that students are to be treated as whatever gender they self-designate. This includes being allowed to use whichever bathrooms or changing rooms they feel most comfortable using. Pronouns are now fluid and children can choose which pronouns are used for them.[3]

2. Mollenkott, *Omnigender*, 20–38.

3. "Alberta Students to Define Their Own Gender."

Mollenkott justifies her vision of gender fluidity by referring to ancient religions whose deities are female and to gender fluidity in various cultures. Alternative genders are part of cultures across the world and included within major religious traditions—Buddhism, Islam, and Hinduism. I mention two of her examples here. The first is the "two-spirit" person in Native American cultures. The details differ from tribe to tribe. Two-spirit persons among Navajos were given sacred status and often acted as shamans, who could cure disease or mental illness. Another significant gender variant is the *hijra* of India. These are castrated men who dress and act like women. They are considered sacred because they have dedicated themselves to Hindu goddesses. Mollenkott claims that Hindus accept the *hijras'* contradictory positions without needing to change it.[4]

Although Mollenkott insists that gender fluidity is positive and accepted by various cultures, the reality appears to be otherwise. For example, the *hijras,* who Mollenkott claims have a positive status in India, are not well liked there, but rather avoided and feared.

> The *hijras* are mainly people with a gender identity disorder or else boys who have been kidnapped and castrated and forced to join the group. Though they wear women's clothes and have female names, they don't look or sound very feminine. . . . In general, their group is viewed with fear and suspicion and is often subjected to ridicule.[5]

The gender fluidity that Mollenkott is so keen to promote is not as well-received in the cultures where it exists as she would have us believe.

As an evangelical Christian, I cannot endorse the idea that we should do away with male and female, nor can I endorse the notion that gender is fluid. But neither should Christians accept everything that has become attached to the labels masculine and feminine. This chapter provides a sound theological basis for upholding two sexes and two corresponding genders. At the same time, we must understand that there is a real place in the kingdom of God for people whose biological sex is ambiguous. In Christ there is room for all people who would trust him, regardless of their sexual biology.

4. Mollenkott, *Omnigender*, 146–76.
5. Warne and Bhatia, "Intersex, East and West," 189.

Gender Wars

Evangelical Christians should be concerned about a real cultural shift towards accepting gender as fluid. This idea finds its ultimate ground in paganism and Gnosticism. Andy Crouch observes that those who lobby for LGBTQ concerns would have us believe that there is no significance in physical sex. The only thing that matters is what the person feels in the heart. Where this view falls down is that it fails to give importance to bodies. Its foundation is Gnosticism[6] and the Gnostic hatred of the physical body.[7]

Peter Jones describes the connection of the gender agenda[8] to paganism. Cultural insistence on unity in humanity and religions is the hallmark of a spirituality of mystical oneness. To attain to this oneness requires that we abandon Christian doctrine, or indeed any doctrine, since these divide. Instead we must look within ourselves. This paganism, dressed up in culturally acceptable garb, desires to eliminate all distinctions between male and female. Sexual identity is now structured around what Jones calls "the monistic ideal of androgyny."[9] According to Jones, in paganism, "The androgyne is the physical symbol of the pagan spiritual goal, which is the merging of two distinct entities, the self and God, and a mystical return to the state of the godhead prior to the mistake of physical creation."[10] Note that when Jones uses the word *androgyne* he is centrally making reference to homosexuality and bisexuality, in which the person plays the role of both the male and the female in the sex act.

We live in a culture that is trying very hard to eliminate the difference between male and female. Christians should not adopt this viewpoint, since this view is based in pagan ideas. Not only is the move to glorify gender fluidity in our culture becoming stronger, but those who advocate for it are using the existence of intersex people as "evidence" for their agenda. Intersex people should not be used as pawns to promote a pagan agenda. Human beings, however sexed, are both valued and dignified, and therefore

---

6. Gnosticism is an ancient religious movement, which claimed that the physical world is evil. The way of salvation in Gnosticism is a particular form of spiritual knowledge, known only to a few.

7. Crouch, "Sex Without Bodies."

8. For more on the gender agenda, see chapter 1.

9. Jones, "Sexual Perversion," 260–63. The quote is found on page 263.

10. Ibid., 271–72.

should not be abused in this way. Intersex people do not exist to justify the abolition of the gender binary.

A Theological Position on Sex and Gender

Christians need a strong theological understanding of human sexuality. This section considers why there are two human sexes. Human sexuality is made to parallel, in some sense, the relationships in the Trinity. Articulating this parallel requires some deep theological reflection on what God has revealed about himself as Father, Son, and Holy Spirit, and some caution, since God is not a physical being. Another reason for human sexuality is marriage. The physical institution of marriage points us to the spiritual marriage of Christ and the church. Therefore, we must not deny the importance of two human sexes.

The Trinity

A theological understanding of sex and gender must begin with the Trinity. Reformer John Calvin observes that knowledge of God and knowledge of humanity are profoundly connected.[11] Because we are made to image God, we would expect that in some way the nature of God would be reflected in humanity. The fact that Genesis 1:27 speaks of the image of God and creation as male and female together suggests a connection between two human sexes and the nature of God. It is wrong to equate humanity as male and female with the image of God,[12] but an examination of God's triunity will shed some light on why humans are made male and female.

In its most basic form the doctrine of the Trinity can be stated as follows. There is one God, who exists as a fellowship of three persons—Father, Son, and Holy Spirit. To put this another way, there are three persons in one God. There is not space here to discuss the foundational evangelical doctrine of the Trinity in depth, only how it pertains to sex and gender. Central to this topic is the fact that Jesus is both one with the Father and yet distinct from him. Jesus said, "I and the Father are one" (John 10:30). Yet passages such as those describing the baptism of Jesus (Matt 3:16–17) demonstrate the real distinction between Jesus and the Father. The Holy

---

11. Calvin, *Institutes.* I.i.1–2

12. For more on this point see chapter 2.

Spirit is also God and distinct from the Father and the Son (Matt 28:19). These are not contradictory statements; they form the basis of the Christian understanding of God.

The Christian tradition in the West mainly stems from the theology of Augustine. He had trouble understanding the distinctions between the three divine persons.[13] Therefore, most Western Christians also have trouble with this concept. On a popular level, most Christians don't think about the Trinity at all. They are confused about how there can be three persons in God. But this is precisely what we must understand if we are to grasp the reason for human sexuality. The concept of person is therefore central.

When we hear the word *person* we tend to think that this equals an individual. But individuals do not need others in order to be themselves. To be a person, as against an individual, is to exist in relationship with other persons. God exists as three persons in relationship with one another. Father, Son, and Holy Spirit are one because their relationship is so intimate and close that they cannot be separated. The three divine persons are not individuals who could disagree or be separated. Yet they are distinct. The Father is distinct from the Son. The Son is a distinct person from the Holy Spirit. They are completely united in loving relationship. But there could be no relationship if there were no distinctions between the persons.

As a communion of persons, the Trinity exists as unity-in-distinction and distinction-in-unity. Unity and distinction are of central importance to a Christian understanding of human sex and gender. Men and women share humanness. This is parallel to the oneness of God. Male and female are distinct in that being male is different to being female. Regardless of the gender roles that may be ascribed to men and women in any given culture, there is a difference between the sexes. This is God's intent, since in some way it parallels the distinction between the trinitarian persons. As Stanley Grenz observes:

> The doctrine of the Trinity makes clear that throughout eternity God is the fellowship of the three persons. No wonder, then, that God's image-bearers best reflect the divine nature in their relationality. The first creation narrative asserts that when God made humankind he built into human existence as male and female the unity in diversity that characterizes the eternal divine reality.[14]

13. Gunton, *The Promise of Trinitarian Theology*, 32–33, 95.
14. Grenz, "Theological Foundations for Male-Female Relationships," 622.

Deeper Reflection on the Trinity

At this point in the argument there seems to be a problem with regard to numbers. How can two human sexes parallel the relationships of three persons in the Trinity? The special role of the Holy Spirit within the Trinity and among human relationships explains why there are not three human sexes.

In his contemplation on the Trinity, twelfth-century Richard of St Victor argued that there must be multiple persons in the Godhead. Since God is good, he must love. In order to love in the most perfect way, there must be another to love and the recipient of the love must be equal to the lover. The love must also be returned or it would not be perfect love. Yet the Christian tradition declares that there are three persons in the Godhead. So Richard argued that perfect love must be shared. Therefore, it cannot be love only between two, but it needs a third person to experience the love of the lover and beloved. Perfect love needs three.[15]

Victor's theology emphasizes the trinitarian persons as those who love. In parallel to God, human beings are created as male and female for the purpose of love. Being male and female is the primary distinction of humanity, although it is not the only distinction between human persons. Therefore, being male and female is the fundamental basis of human love, but not the only one. That is to say, love—mutual self-giving for the benefit of the other—can be expressed and indeed should be expressed between two men or two women. This kind of love is not erotically sexual, but a way of relating to one another which acts for the good of the other.

There are three persons in God and yet there are not three human sexes. This may be explained by a deeper consideration of the Trinity. Christian tradition, beginning with Augustine, calls the Spirit "the bond of love" between Father and Son. In many places the Bible associates the love of God with the Spirit (e.g., Rom 5:5; 1 John 4:12–13). The Holy Spirit is not the divine love as such, but rather the Spirit unites the Father and Son in their mutual love for one another. He enables the trinitarian persons to look outward, to love not only one another but also what is outside of God. The Holy Spirit, as a third divine person, makes the love of the Father and Son greater than it could have been without a third person.[16]

This tradition helps make sense of the fact that there are two human sexes, not three. Just as the Holy Spirit unites the Father and the Son in love

15. Franke, "God is Love," 110.

16. Pinnock, *Flame of Love*, 38–40.

within the Godhead, the Spirit unites believers in the church in love (e.g., Eph 4:3). He indwells believers and thereby ministers the love of God to them and through them (Rom 5:5). The preeminent fruit of the Spirit is love (Gal 5:22), since all the others spring from love. The principal relationship of love is marriage and therefore it is evident that the Holy Spirit binds together husband and wife in love (see Eph 5:25, 28; Col 3:10; Titus 2:4). Indeed, it is not going too far to say that in any relationship where genuine love is present the Holy Spirit has been active.

There are three persons in the Trinity and this must be so, but there are only two human sexes. In the same way that the Spirit binds together the Father and the Son within the Godhead, he binds together male and female in marriage. The bond of marriage is the basis of human community. There are consequently two human sexes and not three. The third person in the Trinity does not necessitate a third sex, since the Spirit plays a parallel role in human life as he does in God. He is the one who unites others in love.

Is There Sexuality in God?

This discussion of the parallels between God's being and human sex and gender begs the question of whether there is sexuality within God. The answer must be both "no" and "yes."

At first we must say that there is no physical sexuality in God. Many religions have female deities. This happens when a society is based on the created structures of human existence and the deity worshipped is a projection of such structures into the supernatural realm. Although the cultures surrounding Israel had female deities and male deities, no such thing was acceptable in Israel. God had to be seen as beyond sexuality, neither a male deity nor a female deity. God is not less than sexual but he is not sexually differentiated in the way that humans are.[17]

The early church fathers made clear that we cannot attribute gender to the persons of the Trinity. During the Arian controversy of the fourth century, the Arians insisted that Jesus was a created being and the Father became Father when he created the Son. The Nicene fathers responded to this claim by stating that we must not understand Father and Son in the Trinity in the same way that we understand father and son in human society. The Father eternally begets the Son and the Son is begotten not made. The Father is not defined by some analogy with human fatherhood

17. Jenson, *The Triune Identity*, 14–15.

or human masculinity, but rather defined by his relationship with Jesus Christ (Matt 11:27).[18]

God is not sexual in the physical sense that humans are, but there are some indications of sexuality in God. Since humans are created in the image of God, there must be something in God that corresponds to sex or gender. The Bible contains many different gendered metaphors for God, mostly masculine, but some feminine. Gendered metaphors along with the marriage relationship, which describes how God desires to relate to his people, demonstrate that in some way God's inner being is like human sexual distinctions. Human community somehow parallels God as a community of persons. These parallels suggest that God is a sexual being, particularly if we see sexuality as something that drives us towards communion with others.[19]

Sexuality as a means of facilitating communion suggests another parallel to human sexuality within the being of God. We would be incorrect to attribute sex to God in terms of *embodiment* as male and female, or to believe that sexual intercourse has a *physical* parallel in the being of God. Yet physical sexual intimacy is grounded in God's triune being. Jesus told his disciples that he and the Father are one (John 17:11). He also promised the disciples that they could share in this oneness (17:21). Indeed sharing in the relationship between the Father and the Son is the goal of the Christian life (17:3). The relationship between Father and Son is one of intense personal intimacy. They know one another without any reserve. Sexual intimacy between husband and wife is also intended to involve an intensity of giving of one person to the other without reservation. So we might say that human physical intimacy in marriage finds its ground in the intense personal intimacy between Father and Son in the life of God. In this way we can say that there is sexuality within God.

Understanding the Beginning from the End

A second way of understanding the importance of human sex and gender is to begin at the end, that is, the goal of human existence. Marriage to Christ is the final destiny of the church. As the final book of the Bible draws near to its end, we are told:

18. Torrance, *Worship*, 88–89.
19. Grenz, "Is God Sexual?," 192–93, 208–11.

> Then I heard what seemed to be the voice of a great multitude, like the roar of many waters and like the sound of mighty peals of thunder, crying out, "Hallelujah! For the Lord our God the Almighty reigns. Let us rejoice and exult and give him the glory, for the marriage of the Lamb has come, and his Bride has made herself ready; it was granted her to clothe herself with fine linen, bright and pure"—for the fine linen is the righteous deeds of the saints. And the angel said to me, "Write this: Blessed are those who are invited to the marriage supper of the Lamb." And he said to me, "These are the true words of God." (Rev 19:6–9)

These final events are prefigured in both Old and New Testaments. The prophets in the Old Testament spoke in different ways about the relationship between Yahweh and Israel. A common metaphor was that of marriage, with Yahweh calling himself the husband of Israel. This motif is repeated in several prophetic books. "For your Maker is your husband, the LORD of hosts is his name; and the Holy One of Israel is your Redeemer, the God of the whole earth he is called" (Isa 54:5). Allusions to Yahweh as Israel's husband are made in Jeremiah 2:32 and 3:1, and more explicit claims to this are found in Jeremiah 3:14, 20 and 31:32. The prophet Hosea is well-known for marrying a woman of prostitution as a parable of the relationship between Yahweh and his people (Hos 1:2). Yahweh promises, "And in that day, declares the LORD, you will call me 'My Husband,' and no longer will you call me 'My Baal'" (2:16).

The New Testament also uses marriage as a significant metaphor for the relationship between God and his people. The marriage metaphor is most explicit in the letter to the Ephesians. There the Apostle Paul writes:

> Wives, submit to your own husbands, as to the Lord. For the husband is the head of the wife even as Christ is the head of the church, his body, and is himself its Savior. Now as the church submits to Christ, so also wives should submit in everything to their husbands . . . "Therefore a man shall leave his father and mother and hold fast to his wife, and the two shall become one flesh." This mystery is profound, and I am saying that it refers to Christ and the church. (Eph 5:22–23, 31–32)

In this important passage regarding the relationship between husbands and wives, Paul reveals that the submission of wife to husband, and the sacrificial love of husband for wife, is based on the relationship between Christ and his church. The statement, "Therefore a man shall leave his father and mother and hold fast to his wife, and the two shall become one flesh," is taken from Genesis 2:24 and was first written about Adam and Eve. Yet Paul concludes that it applies to Christ and the church. The upshot of verse 32—"This mystery is profound, and I am saying that it refers to Christ and the church"—is that the original human couple was created for more than procreation or as the basis for human community. Rather, the first marriage, and every marriage that followed it, is sacramental in that it points to a spiritual reality beyond itself.

Therefore, marriage between a man and a woman is very significant. It is an important reason for two human sexes. Without being sexed and gendered beings, humans cannot live out the sacrament of marriage, which points us towards our destiny. This of course raises some questions about people who are single and who never marry. Not to mention those issues that it presents for people who are intersex. These matters are addressed in more depth later in the chapter. For now, let it be said that although marriage is a very significant sign of the final state of humanity, this does not mean that people who do not marry cannot also participate in that final state. Indeed, being single is a profound New Testament option for Christian discipleship.

Salvation, Gender Roles, Discipleship, and Intersex

It is clear that humans are intended to be physically sexed and socially gendered beings. If the existence of two distinct sexes is so theologically important, then where does this leave the person who is intersex? Being ambiguously sexed is not a hindrance to being part of God's kingdom. Intersex people are brought into the kingdom and become disciples of Jesus in the same way as the unambiguously sexed. Salvation, gifting, and discipleship are not dependent on gender.

Jesus: Savior of Humans, However Sexed

Jesus is a male human being. Many feminists believe that this is a problem, because they insist that if Jesus is male then God is male and women are thereby marginalized. But these feminists have misunderstood the significance of Jesus and projected human qualities onto God. Jesus' male body does not mean that he can only be the Savior of males. He is the Savior of humans, however sexed.

The New Testament assumes that Jesus is biologically male. He was circumcised on the eighth day, and his parents offered the sacrifice for the firstborn male (Luke 2:21–23; see Exod 13:2, 12), so he must have appeared convincingly male. But the maleness of Jesus is a problem for some feminist theologians. Some complain that because of a male Christ feminine language about God is ignored or rejected. The male Jesus is the ideal human, thus making male humanity closer to the ideal than female humanity. Women are thereby marginalized.[20] Others suggest that women cannot be saved by a male Savior, but need a female Savior to be empowered. They claim that a male Christ is the justification for all patriarchal structures in society.[21]

An immediate response to the feminist insistence that a male Savior is offensive and excludes women is needed. First, the early church developed the doctrine of *anhypostasis*—from the Greek *an*, "not," and *hypostasis*, "person," meaning "no person"—to describe the idea that the Word became flesh (John 1:14). The person of Jesus Christ exists because of the union of the Son of God—the eternal Word—to humanity, but the human nature of Jesus did not and does not exist apart from the incarnation. The human nature of Jesus came into existence only as the Word took on flesh.[22] This implies that the Son of God—the eternal Word—is not a male. The person of Jesus is a male human being. But we cannot project the maleness of Jesus back into the Godhead as if the human nature of Jesus is male because of the maleness of God. This is moving in the wrong direction and misunderstands the incarnation.

Second, the feminist complaint is an example of what is known as the "scandal of particularity," that is, the offense that God has manifested himself in a particular human being from a particular nation in a particular time in history and no other. Jesus Christ is the revelation of God to humanity and he happens to be male. Since God is revealed through this particular human being, we have no right to decide that we would like this to be otherwise. Insisting that each person must have a Savior of their own gender is a symptom of the human tendency to project human qualities onto God. It would be fruitless to insist that we need an androgynous Savior, because Jesus could then only save people who are intersex.

---

20. Kim, "Revisioning Christ," 82.

21. Green, "More Musings on Maleness," 9–12.

22. Torrance, *Incarnation*, 67, 84, 228–29.

Jesus is a male human being, but this does not imply that he can only represent or be the Savior of male humans. The New Testament emphasizes the humanity of Jesus far more than it speaks of his maleness. If we exclude references to Jesus as Son and the use of the personal pronoun "he," there are surprisingly few mentions in the New Testament to Jesus' maleness. There are only three instances in the New Testament that specify Jesus as an adult male using the Greek word *anēr*—Luke 24:19,[23] Acts 2:22, and 2 Corinthians 11:2. On the other hand, Jesus is frequently referred to using the generally gender-neutral noun *anthrōpos*.[24] *Anthrōpos* refers to a human being, rather than a person of a particular gender, although translations often obscure the gender neutrality of the term.

Jesus often refers to himself as the Son of Man.[25] The Greek word for man in this expression is not *anēr* (male human) but *anthrōpos* (human being). Although "Son of Man" signifies more than Jesus' humanity, it nonetheless designates Jesus as a human being, without being specific about gender. There are a number of other places where Jesus is designated by the gender neutral expression *anthrōpos*. Passages that refer to Jesus as *anthrōpos* include: Luke 4:4 (indirectly); John 4:29, 9:11, 19:5; and Hebrews 5:1 (indirectly, see 5:5). Two directly pertinent passages are Romans 5:15 and 1 Timothy 2:5. Romans 5:15 tells us that both Adam and Jesus represent all humanity. Those who are in Adam die, regardless of their sex. Those who are in Christ live, again regardless of sex. It is not as a male that Jesus mediates between God and humanity (1 Tim 2:5), but as a fellow human being, a unique human being who stands on both sides of the divide—human and divine. In this way Jesus can be the Savior of all who would put their trust in him. His gender does not prevent him being Savior to men, women, and intersex.

Male and Female Roles

For some Christians, gender roles in church, family, and society are very important. Certain views on gender roles could present a problem for people with ambiguous sex, because men and women must perform different

23. Most English translations do not translate the word *anēr* in this verse.

24. The exceptions are Matthew 19:5, 10 and 1 Corinthians 7:1, in which it refers only to adult males.

25. Matthew 8:20; 16:13; 17:19, 22; 19:28; 26:64; Mark 8:27; 14:62; Luke 9:58; 22:48; John 6:53; 8:28; 9:35; 12:23, 34; 13:31.

roles. Depending on how we view gender roles, intersex persons will either have difficulty finding a place or they will experience freedom to express their uniqueness. I will explore two major positions here: complementarian and egalitarian. I will advocate for a more egalitarian position, particularly in church. The egalitarian position would provide no hindrance to intersex persons taking up any role in church or society.

Two significant views of gender roles exist among evangelical Christians at this point in history. The first is known as complementarianism. Some of the greatest champions of this view are to be found in the Council for Biblical Manhood and Womanhood, who affirm, "Distinctions in masculine and feminine roles are ordained by God as part of the created order, and should find an echo in every human heart."[26] Men and women were created with much in common: they are both in God's image, they share equally in salvation, and both are to rule over the earth through procreation. However, the creation narrative in Genesis 2 makes clear that men and women are significantly different. There is a "hierarchical ordering of roles between the first man and woman."[27] Since Adam and Eve are the model for all subsequent marriages, men must lead and women must follow. Although women are equal in humanity they are subordinate in role. The New Testament assigns different roles to men and women in home and church. Women are to submit to their husbands. In the church the leadership roles are likewise restricted to men.[28]

There are some significant issues with the complementarian view. The first of these is a theological one. Complementarians justify their position by asserting that the Son is subordinate to the Father within the Trinity, not merely during Jesus' time on earth. Conservative evangelicals who take this stance insist that it is not to be confused with the heresy of subordinationism, which was adopted by Arians in the fourth century. Arius taught that Jesus was subordinate to the Father, both in what he does and in his essence. Complementarians claim that Jesus is truly equal with the Father in essence, but eternally subordinate to him in role, by virtue of what it means to be Son. The problem is that it is not enough to simply uphold the oneness of the divine essence. If the Son is eternally subordinate to the Father, then there is a hierarchy of deity in the Godhead,[29] a clearly heretical position.

26. "Danvers Statement."

27. Burk, *What Is the Meaning of Sex?*, 163.

28. Ibid., 163–69.

29. Giles, *The Trinity and Subordinationism*, 23–27; Giles, "Barth and

The second issue is about interpretation of the creation narrative. The debate concerns the meaning of the word "helper" in Genesis 2:18. Complementarians believe that Eve was intended to be Adam's assistant, some suggesting that she helped him by taking care of domestic chores while he farmed.[30] This makes the help that Eve provided to Adam very narrow in scope, confined to the domestic sphere. Yet the Hebrew word in view here is certainly far greater in scope than domestic help. "Helper" is frequently used to describe the God of Israel. Yahweh helps Israel in battle. The Psalms tell of Yahweh helping those in need—the poor, the sick, orphans, and the oppressed. Thirteen times "helper" refers to God's ability to save or deliver his people.[31] Therefore, interpreting Genesis 2:18 to mean that Eve was to be Adam's assistant is quite inappropriate. Eve's help to Adam is far more significant than mere domestic assistance. The second-century theologian Irenaeus is closer to the mark in attributing to Eve the task of helping the immature Adam grow to perfection.[32]

In contrast to the Council for Biblical Manhood and Womanhood, which emphasizes the differences between men and women, Christians for Biblical Equality stresses the unity, common purpose, salvation, and joint responsibility of men and women. They seek to read the Scripture holistically, rather than concentrating on a few select passages that seem to restrict the roles of women. Genesis teaches that men and women share the image of God, the responsibility for bearing and rearing children, and the call to have to dominion over the earth. The New Testament teaches that men and women both gain salvation through Christ, both are indwelt by the Holy Spirit, are given gifts of the Spirit without reference to gender, and are called to prophetic, priestly, and royal duties given through Christ. Since Christian ministry is service, not domination of another, both men and women can lead in both church and home. Husbands and wives are to defer to one another and to use their gifts for the benefit of the household. In this way both men and women are able to live out their freedom in Christ, living according to gifting rather than gender restrictions.[33]

Clearly the complementarian position would prove more difficult for people who are intersex, because without being sure of whether you are

Subordinationism," 328–31.

30. Gellman, "Gender and Sexuality in the Garden of Eden," 330.

31. Harman, "6468 עזר."

32. Steenberg, *Irenaeus on Creation*, 151.

33. "Men, Women and Biblical Equality."

male or female you would not be sure which role to fulfill. The egalitarian position is not merely better for women, many of whom are frustrated by the restrictions that complementarianism places on the use of their God-given gifts, it is also less problematic for intersex persons. Instead of needing to be sure of what sex you are in order to decide on role or restriction of role, intersex persons can use the gifts they have been given as human beings for the benefit of other human beings. Intersex people are well able to express their interdependence with other humans. Being intersex does not diminish the need for service to others or for being served by others. Nor should being intersex stand in the way of using natural talents or spiritual gifts. If God does not distinguish the outpouring of his grace on the basis of gender, then gender roles need not get in the way of intersex persons living out grace-filled and grace-empowered lives. This applies most simply in social, public, and church roles. There are still problems which relate to marriage, since who you marry or avoid marrying is not solved by the egalitarian position. That issue will be considered in chapter 6.

Being Like Christ

Being sexed and gendered beings is important and we should continue to uphold the ideal that humans are intended to be male and female. However, being intersex should not prevent a person attaining the goal of human existence, which is to become like Jesus (Rom 8:29). Although men, women, and intersex persons will each express themselves in unique ways, those who follow Jesus are all seeking to acquire the same character qualities. Eastern Orthodox writer Nonna Verna Harrison observes:

> Gender stereotypes, which assign some virtues to men and others to women . . . introduce inappropriate fragmentation, disrupting the human wholeness and unity for which God created us. If Christ lives in a man, so does the Holy Spirit, and if the Spirit lives in a woman, so does Christ. Likewise, as the Fathers say repeatedly, to exercise any one virtue perfectly, be it courage, nurturing, or whatever, requires all the virtues.[34]

There are three angles from which to look at gender and likeness to Christ: through the qualifications of a disciple, through the actions of Jesus, and through virtues. There are both male and female disciples described

34. Harrison, "Orthodox Arguments against the Ordination of Women as Priests," 173.

in the Gospels. Being a disciple of Jesus, then, is not restricted by gender; discipleship for intersex people is a genuine option. Jesus is male, but he also fulfilled roles that are frequently ascribed to females. Thus being like Jesus does not require a person to be a particular gender. The virtues that the New Testament exhorts Christians to aspire to are applicable to both men and women. There are, then, no virtues that require a person to have a particular gender.

We may be accustomed to thinking of the disciples of Jesus as male. But there are female disciples and these are portrayed very positively by the Gospel writers. The Gospel of Mark describes Mary Magdalene, Mary, and Salome as people who followed Jesus (Mark 15:40–41), that is, they were disciples. Women are mainly shown as model followers of Jesus. Simon's mother-in-law ministered to Jesus (1:30–31). The woman with the issue of blood had faith (5:34). Jesus commended a widow as an example of extravagant giving (12:42–43). A woman anointed Jesus in Bethany for his impending death (14:3–9). Women were first to see the empty tomb and they were given the task of proclaiming the resurrection (16:1–7). Women are included by Jesus within his family as defined by those who obey the will of God (3:35). By the same token, they will be the recipients of persecution for following him (10:30). According to Mark's understanding of discipleship—that is, being a disciple is a matter of following, serving, and suffering—women are prime examples.[35]

The point we should take from this is that being female is not a hindrance to being a faithful disciple of Jesus. By extension, being intersex is not a hindrance to being a faithful disciple of Jesus. Nothing about being intersex will intrinsically prevent someone from following Jesus, serving him, and suffering for his sake. The only qualification for being a disciple of Christ is faith. "Whoever believes in him [Jesus] is not condemned, but whoever does not believe is condemned already, because he has not believed in the name of the only Son of God" (John 3:18).

The second way of considering how becoming like Jesus is not dependent on gender is to look at the actions of Jesus. Jesus is clearly male and he exhibited the required traits to be considered masculine in the ancient Mediterranean world. Significant masculine traits that Jesus possessed include self-mastery and authority over others. In the ancient world, self-mastery for males was demonstrated particularly through control over sexual passions. Jesus also plainly demonstrated authority over others

35. Thurston, *Women in the New Testament*, 68–69, 77.

throughout his life. In addition, his masculinity was evidenced by courage, living justly, expressing no need of luxury, and anger over injustice.[36]

There are, however, some actions of Jesus that we might associate with women more than men. Jesus exhorted his disciples to be servants of others (Mark 9:35) and told them that he had come not to be served but to serve (Matt 20:28). Jesus served the disciples by washing their feet (John 13:3–5), a task generally assigned to slaves. Jesus also drew children and infants to himself. His disciples were outraged by this behavior (Mark 10:13–15). He was financially supported by others (Luke 8:2–3) and owned no property of his own (Luke 9:58). Jesus cooked breakfast for his disciples after his resurrection (John 21:9). Jesus' testimony was ignored (John 8:12–18), paralleling the ignored testimony of women in first-century Judaism. Like a woman, Jesus had all decisions taken out of his hands when he was falsely accused and executed. Lastly, Colleen Conway observes that although Jesus is clearly male and clearly masculine in his relationship with other humans, "when it comes to his relationship with God the Father, he assumes a less masculine status. He is obedient, submissive, can do nothing on his own (John 5:19, 30; 6:38; 7:16; 12:49; 15:10)."[37]

Despite the fact that Jesus is both biologically male and genuinely masculine by first-century standards, he engaged in activities that are often considered feminine and shared experiences that intersect with the experiences of many women. Therefore, the maleness of Jesus does not imply that becoming like him is possible only for men. If Jesus can act in ways that are masculine and ways that might be considered feminine, then those who are intersex will not be prevented from becoming like Jesus.

Lastly, the virtues exhorted by the Bible are for both men and women, and thus for those who are intersex. Most of the exhortations in the epistles are directed at every disciple and are not gender specific. In the early church, both men and women were persecuted and imprisoned, possibly even killed, for their faith (Acts 9:1–2; 22:4). This suggests that women as well as men need to be strong and courageous to live as disciples of Christ. By the same token, men must be gentle (Eph 4:2; 1 Tim 3:3) as Jesus is gentle (Matt 11:29), even though gentleness is a seemingly feminine quality. Even where the Bible exhorts men and women to different roles, as in marriage, the character qualities required are the same. The classic passage about husbands and wives, Ephesians 5:22–33, exhorts wives to submit to

36. Conway, "'Behold the Man,'" 166.

37. Ibid., 179.

their husbands and husbands to love their wives. This is preceded by the command, "submitting to one another out of reverence to Christ" (5:21). Both husbands and wives are to submit and to express love. What differs is the means by which this is done. Wives express love by submitting to husbands; husbands submit to their wives by loving them.

There are no separate spheres of discipleship for women and men. Therefore, being an intersex person cannot prevent someone from being a disciple of Jesus. No one needs to be concerned about discerning which sphere to position oneself in before becoming a disciple of Jesus. The person who is intersex can be transformed into the likeness of Christ without the need to be sure about sex or gender. Transformation into the likeness of Jesus depends not on gender, but on union with the Son through the indwelling of the Holy Spirit. This, then, may be the upshot of Galatians 3:28—"There is neither Jew nor Greek, there is neither slave nor free, there is no male and female, for you are all one in Christ Jesus"—for intersex persons.

Intersex and Gender

There is nothing intrinsic about being intersex which prevents a person following Jesus, using his or her gifts and talents in church and society, or becoming a person of character. This is good news. Yet the world is still dominated by male and female. Is there a positive place for intersex people in the kingdom of God? The Bible does provide a positive place for people who are unusually sexed, particularly in Jesus' commendation of the eunuch. What about marriage? If marriage is the destiny of humans, can intersex people, who may be unable to marry, still be a sign pointing to the wedding supper of the Lamb? Celibate singleness is also upheld by the New Testament as signifying the importance of marriage. Marriage is of course not the only relationship in life and intersex persons, as with unambiguously sexed persons, are not defined by their genitals. This chapter concludes with a final statement about intersex and gender.

The Eunuch

Although the Bible is clear that God intended humans to be male and female, there is good news for those who fall outside the sexual binary. The biblical category of the eunuch suggests a clear place for intersex persons within the church and the kingdom of God. There are two mentions of the eunuch in the New Testament. Both offer positives for intersex persons. The first is a saying of Jesus and the second an incident in the book of Acts. Jesus said,

> For there are eunuchs who have been so from birth, and there are eunuchs who have been made eunuchs by men, and there are eunuchs who have made themselves eunuchs for the sake of the kingdom of heaven. Let the one who is able to receive this receive it. (Matt 19:12)

The context of this saying is a discussion of marriage and divorce (19:3–11). Jesus affirmed that God made humans as male and female (19:4), yet at the same time he affirmed the eunuch. Jesus listed three kinds of eunuchs: those who are eunuchs from birth, those made that way by others, and those who make themselves eunuchs for the sake of the kingdom of heaven. The third category has been interpreted in different ways in the history of the church, with some men castrating themselves so that they could become literal eunuchs for the sake of the kingdom of God, and others merely adopting a life of celibacy without the genital mutilation.[38]

The first category of eunuch is closest to intersex. In Hebrew "eunuchs who have been so from birth" are *saris khama,* eunuchs of the sun. These were likely people with intersex conditions. There were three categories in rabbinic literature of people with unusual sexual organs: *aylonith, androginos,* and *tumtum. Aylonith* are women with underdeveloped genitalia and no pubic hair. They are unable to engage in penetrative sex and are infertile. *Androginos*, that is, hermaphrodites, are mentioned in the Talmud. They are considered to be both male and female and must conform to the laws and obligations of both. The third category is *tumtum*, which means "sealed over," because it is very difficult to tell at birth which sex is applicable. The rabbis knew that God created humans as male and female, but acknowledged people who did not fit either category. They developed laws that spelled out the rights and obligations of people who were not clearly male or clearly female. In this way the *saris khama* had a place within Jewish society, and with this came obligations and rights as well as protection.[39]

The second category, the castrated eunuch, was often despised in ancient Israel and the ancient Mediterranean world. In Jewish law there is a

---

38. DeFranza, *Sex Difference in Christian Theology,* 72–73.
39. Hare, "Hermaphrodites, Eunuchs, and Intersex People," 83–87.

prohibition against castration (Lev 21:20; 22:24; Deut 23:1). The eunuch was stigmatized because of his inability to father children, something that was expected of all males.[40] Eunuchs were feared by men because of their status as not quite male and not female. There were, it was assumed, no virtuous eunuchs.[41] For this reason it is extremely interesting that Jesus spoke so positively of eunuchs. Instead of physical offspring, the eunuch is lauded as one who can produce spiritual offspring. The eunuch saying is set within the wider context of sayings which elevate the value of being like a child (19:13–15) and giving up material possessions (19:16–30). These sayings, too, are countercultural and adopt a position that goes against the gender expectations for men. The kingdom of God is advanced by those who choose to adopt the social status of lesser males or the even lower status of eunuchs, children, and slaves.[42]

The New Testament elsewhere ascribes positive status to a eunuch. In Acts 8:26–40, Philip met an Ethiopian eunuch, who was reading about the death of the suffering servant (Isa 53). On the same scroll a significant promise regarding eunuchs is found.

> For thus says the LORD: "To the eunuchs who keep my Sabbaths, who choose the things that please me and hold fast my covenant, I will give in my house and within my walls a monument and a name better than sons and daughters; I will give them an everlasting name that shall not be cut off" (Isa 56:4–5).

This passage looks forward to something new, an inclusive worshipping community in which eunuchs, foreigners, and others who have been excluded, can now be included if they faithfully keep the Sabbath and are obedient to the covenant. The suffering of the servant in chapter 53 is followed by a reference to his resurrection from death. Thus the salvation of the Ethiopian eunuch is prophesied there. The eunuch, humiliated by his place within that culture, would find a place of inclusion within the new community inaugurated by the risen suffering servant, Jesus Christ.[43]

Although the eunuchs from birth that Jesus mentioned in Matthew 19:12 are likely intersex persons, we cannot simply equate the biblical eunuch with intersex. The eunuch in Acts 8 was probably once an undisputed male who was castrated, and thus not intersex. However, consideration of

40. Melcher, "A Tale of Two Eunuchs," 118.

41. DeFranza, Sex Difference in Christian Theology, 76–77, 79.

42. Anderson and Moore, "Matthew and Masculinity," 90–91.

43. Melcher, "A Tale of Two Eunuchs," 118–24.

the biblical eunuch opens a positive place for intersex persons within the community of God. First, there is a clear place for people with unusual sexual anatomy in the kingdom of God. Jesus does not denigrate or insult, but rather upholds the eunuch as a model of someone who may advance the kingdom of God. Second, the only criterion for being included in the kingdom is faith in Christ, not anything related to physical anatomy. The eunuch in Acts 8 became part of the community of God by believing Philip's preaching about Jesus (Acts 8:35). Therefore, people who are intersex should in no way be hindered from participating in the worshipping community of the church, since it is faithfulness to the covenant through Christ that counts and not sexual anatomy.

Intersex and Marriage

Does the theological significance of marriage make those intersex people who cannot marry second-class citizens in the kingdom of God? The answer is a categorical "no." There is real value placed on singleness in the New Testament. Being single is a valid way of being human. Singleness, like marriage, can point towards the end goal of human existence—union with Jesus. So whether intersex persons marry or not, they are not diminished as people.

Many people who are intersex have married. There are, however, reasons why intersex persons may find themselves unable to marry. Some chromosomal males have had genital surgery as infants to make them into females, something that happens often in the case of infants with micropenis. These and others may find themselves anatomically unable to engage in penetrative sex. For other people, gender confusion or shame may prevent marriage. Some intersex individuals have no sex drive. Infertility might present a hindrance to marrying for some. Legal issues may also impede marriage. There are some legal precedents that mean that, in Australia, marriage to an intersex person may not be legally valid.[44] Then again, intersex persons may find themselves unmarried for a host of other reasons that pertain to every other single person.

If marriage is so important theologically, then what is to become of the intersex person who cannot marry? Should Christians be promoting genital surgery to make intersex persons marriageable? Making intersex people marriageable has been one of the major justifications for genital

44. Wilson, "Intersex People and Marriage."

surgery on intersex infants.[45] However, there is no need to surgically alter infants based on a Christian understanding of marriage. Intersex bodies are acceptable without surgery, and it is quite acceptable for people, intersex or otherwise, to be unmarried. Singleness is a valid expression of human being.

In the Old Testament, marriage was expected of every person, with only a few single people, such as Jeremiah (Jer 16:2), used by God in a significant way. In first-century Jewish society it was expected that all men would marry and father children. But the New Testament gives a more positive view of singleness. Jesus' saying about the eunuch in Matthew 19:12 implies that unmarried people are not inferior to married people; they are both able to advance the kingdom of God. The Apostle Paul also viewed singleness as a positive option and even an advantageous state, because the distractions of marriage and family diminish human capacity to serve God. Therefore, both marriage and singleness are valid states and positions from which to serve God and to be transformed while awaiting the resurrection.[46]

There are two matters that need to be addressed in relation to the single state. First, if marriage is the goal of human existence, then are married people closer to salvation than single people? The New Testament is clear that the way of salvation is through the person and work of Christ, received by faith. Those who will share in the wedding supper of the Lamb are those who have followed Jesus and those who confess him as Lord. Many followers of Jesus throughout the history of the church have been single. Jesus himself was single. Marriage is a means by which God shows us something of what is to come in the union of Christ and the church. Singleness is another means by which God demonstrates the importance of that union. But neither marriage nor singleness is a means to salvation.

Second, how does singleness point to the marriage of Christ and his church? Jesus was unmarried when he walked the earth, because he was saving himself for his bride, the church. In the same way, celibate singles can demonstrate by their purity that there is something far more important to come than sexual intercourse. Paul uses the analogy of a pure virgin to describe the church. "For I feel a divine jealousy for you, since I betrothed you to one husband, to present you as a pure virgin to Christ" (2 Cor 11:2). Married people may be a picture of Christ and the church, but celibate

45. Reis, *Bodies in Doubt*, 56.
46. Hsu, *The Single Issue*, 30–36.

singles are a picture of the purity of the church awaiting her bridegroom. Being married or being single are both means by which God shows in living pictorial form the importance of the eschatological wedding supper of the Lamb (Rev 19:9). Both marriage and singleness may be considered as gifts from God (1 Cor 7:7). Therefore, the celibate single intersex person can point to the glory of the wedding supper of the Lamb, just as much as those who are called to be married.

Intersex and Other Relationships

Marriage is a significant relationship in life, but it is by no means the only significant relationship. Stanley Grenz argues that Adam's sexual nature drew him to seek another human being to break his solitude. Because humans are sexual beings they are incomplete without others. Therefore, we exit our isolation and enter into community.[47] Does someone require an unambiguous sex before the need for community is experienced? Certainly not. Human beings, whether they be female, male, or intersex, need other human beings. We need to enter into relationships with others in order to be whole people. Those relationships do not necessarily need to be marriage relationships. Even married people need relationships beyond the marriage, albeit those relationships are of a different nature.

Indeed most of the relationships in which we engage in life are not marital ones. Men relate to other men and women relate to women. Both men and women relate to one another as men and women on many levels outside of marriage. Men can be fathers or brothers or sons. Women can be mothers or sisters or daughters. People relate to one another as friends, and in the body of Christ we relate to one another as people who share a common Father in heaven and a common brother in the Lord Jesus Christ. As Stanton Jones and Mark Yarhouse explain:

> Sexuality is not the only thing, or the most important thing, about what it means to be a person. . . . Human nature is thus basically and intrinsically sexual, and yet not reducible to, or explicable only in terms of, the various facets of our sexuality.[48]

We are sexual beings, but it is depersonalizing to reduce a man to a penis and woman to a vagina, as if our genitalia were the only part of sexual

47. Grenz, "Theological Foundations for Male-Female Relationships," 621.
48. Jones and Yarhouse, "Anthropology, Sexuality, and Sexual Ethics," 119.

interactions with one another. Humans relate to one another as whole persons, not as parts of themselves. Even in marriage, relationships involve much more than sexual intercourse and sexual anatomy. Intersex persons are not reducible to their genitalia any more than is any other human being. Each enters into a variety of relationships as a whole person. Any given person, intersex or otherwise, has a unique set of characteristics—physical, social, intellectual, and emotional—that make that person uniquely different to others and therefore able to enter into relationships with others. Relationships, not genitals, make us human persons.

Conclusion: Intersex and Gender

Finally, since this is a chapter about sex and gender it is inevitable that we ask the question regarding the sex or gender of the intersex person. Should the intersex person adopt a sex of male or female, or is the eunuch discussed above a precedent for having a third sex? This is not a simple question to answer. I have not ventured to answer this question before this point, because I first wanted to establish a biblical foundation for the value of the intersex person as they are. Intersex individuals are created in love by God and completely acceptable as they are, without alteration. There is a positive place for intersex persons within the kingdom of God.

With these things in mind, I humbly suggest the following. Humanity was created as male and female. There is no actual third sex or third gender in a lasting sense. The eunuch, lauded by Jesus as significant in the kingdom of God, is not a third sex. Intersex should not, then, be considered a third sex, but a physical distortion of male and female. Yet this does not mean that the intersex individual has to choose one or the other in order to be a person. Grace does not require someone to change to be acceptable. The only definitive reason that would require an intersex person to adopt male or female is when entering into marriage. This is discussed in chapter 6.

However, pragmatically speaking, it is possible and helpful to live as either male or female. Life is less complicated if it is lived as a particular gender. Surgical alteration is not required to make this happen. Grace is always available from God in making a decision about which gender to choose. Abundant grace also needs to be exercised by the body of Christ towards intersex people, who sometimes struggle with these decisions. Prayer about this decision is always appropriate. Because this can be a difficult decision for some people, the first choice may not be the best one. Transitioning is an acceptable option, as long as this does not happen within an existing marriage.

The fact that intersex is not a genuine third sex does not in any way make the intersex person inferior to the unambiguously sexed person. Christian believers, therefore, need to consciously refuse to exclude people based on their biology. Since we all need to be in relationship in order to be ourselves, each person—male, female, or intersex—must deliberately make relationships with those on the margins a priority. This will both enhance the humanity of intersex people and the humanity of the unambiguously sexed. In the next chapter I will consider the ways in which Jesus treated people who are on the margins. This will reinforce the value of all people, however sexed, and help Christians to see the need for inclusion of those who are unusual.

4

Jesus and Intersex

Introduction

THOSE WHO ARE MARGINALIZED and rejected sometimes cry out to God to look down from heaven and see the injustice and the pain. If only he "would rend the heavens and come down" (Isa 64:1). In the incarnation, the Son of God did precisely that. The presence of the Son of God, enfleshed as a human being, brought God's justice and mercy to the earth in physical form. Thus the incarnation is good news for intersex people.

In becoming human, Jesus identified with many of the experiences common to intersex people. Jesus was a man with little status. He lived during a time of political and social oppression. Born into poverty and obscurity, he experienced a perilous childhood as a refugee, in hiding from those who sought to kill him. His childhood playmates mocked him as illegitimate. In his ministry he demonstrated the grace of God to the marginalized. This was personally costly, because he identified with those who were rejected and despised. To the sick and marginalized, Jesus gave love and dignity. For sinners in need of salvation, he offered grace and mercy. Although no intersex persons were among those who received ministry from Jesus, there is no doubt about how Jesus would treat intersex people.

God Rends the Heavens and Comes Down

In the person of Christ, God did rend the heavens and come down. But the Son of God did not enter the world in the way we might expect. He did not come to enjoy privilege and power, but chose to identify with the lives of the poor, the weak, and the oppressed. Jesus was not powerful; he had the lowly status of a slave. He inhabited a nation that was politically and religiously oppressed. He was born poor and was subjected to violence and exile from his home as an infant. His childhood was that of a stigmatized individual. Because he experienced all these things, Jesus did not need to be intersex in order to understand the powerlessness, oppression, vulnerability, and stigma of being intersex. The reasons for his experience were not the same, but the pain of the experience is fundamentally the same.

A Man without Status

"The Word became flesh and dwelt among us" (John 1:14a). Jesus came to bring God's justice to the world and he did this first of all by willingly taking a great step downwards, by giving up his glorious position in heaven. A hymnic passage in Philippians tells us how much Jesus humbled himself for our sakes.

> Have this mind among yourselves, which is yours in Christ Jesus, who, though he was in the form of God, did not count equality with God a thing to be grasped, but emptied [kenoō] himself, by taking the form of a servant [doulos], being born in the likeness of men. And being found in human form, he humbled himself by becoming obedient to the point of death, even death on a cross. (Phil 2:5–8)

The word kenoō (v. 7) may be translated as "made himself of no reputation," which means that Son of God laid no claim to his glory. Alternatively we could say he "emptied" himself. This does not imply that the Son emptied something out of himself; he did not empty divinity out of himself. Rather he emptied himself, that is, he gave up his glory and assumed a form that was not glorious, the form of a slave. Even as a man Jesus humbled himself, by submitting to the constraints of being human, submitting to the will of the Father, and dying the death of a slave on a cross. The humility of the Son of God was matched by the humility of the man Jesus.[1]

The Son of God became human and took the form of a doulos—a bond-servant or slave (v. 7). The status of a slave in the ancient world was so low that one might say that a slave had no status at all. Slaves had no

---

1. Torrance, Incarnation, 74–75.

rights. A *doulos* had no possessions of his own; he did not even own his clothes. His master owned everything of his, even his life. The Son of God deprived himself of all rights. Jesus owned clothes, but owned no place to lay his head (Mark 8:20). He possessed nothing large enough that it could not be carried. He borrowed a donkey to enter Jerusalem on Palm Sunday. He was buried in someone else's tomb. Whatever he did was not for his own benefit, but for the benefit of others. He went from overwhelming glory to a life of complete humility.[2] The contrast between Adam and Christ is clear. Adam became a slave to decay and death because of his sin. But Christ, the last Adam (Rom 5:12–21; 1 Cor 15:21–22, 44–49), *chose* to become a slave and live under the consequences of the first Adam's choice.[3]

In a world in which male and female prevails, the status of intersex persons is often doubtful. Since intersex is frequently ignored, or more likely intentionally covered up by surgery and secrecy, from one standpoint intersex has for a long time had no status. What the *kenosis*, or emptying, of the Son of God means first of all is that he chose to share the status of those without status. He chose to identify himself with the people who have no rights, because they have been denied those rights, just as so many intersex people have been denied their rights. He chose to be vulnerable to hurt, pain, and rejection. When the Son of God became the man Jesus, he did so knowing that he would not be in control of his own life. In this way, becoming incarnate was the first step in identifying with intersex people.

Oppression and Freedom

Jesus' identification with intersex people included living under oppression. The nation of Israel was under political domination and religious discouragement. By the time of Jesus, Israel had been under foreign rule for centuries. Following the exile to Babylon, Israel was ruled by Persians and then Greeks. The Romans took control of Jerusalem in 63 BC. Judea was allowed little territory and was ruled by a Roman governor. In 40 BC, Herod the Great became a Jewish client-king under the Romans. He was a ruthless man, willing to assassinate those who got in his way, including his own family members. After Herod died in 4 BC, Judea was ruled directly

---

2. O'Brien, *The Epistle to the Philippians*, 222; MacArthur, *Philippians*, 127.

3. Cousar, *Philippians and Philemon*, 53.

by the Romans, who collected taxes from the Jews to benefit Rome, ignored the local customs, and had frequent conflicts with the people of Judea.[4]

The Jewish mind-set in the first century was shaped by the exile in Babylon. The presence of Gentile rulers indicated that the exile had not truly ended. The great scriptural promises of return from exile remained unfulfilled. Therefore, in the minds of first-century Jews, God was still punishing them for their idolatry.[5] In addition to this state of despondency, the people were under religious oppression from within their own religious establishment (see Matt 23:1–4). The client-king Herod the Great controlled the temple and high priesthood for his own ends.[6] The priests and scribes, as Judea's elite, were allied to Herod, rather than being concerned with the welfare of the people.[7] It is no surprise, then, that the Jews looked for a deliverer and a new exodus.

A world dominated by oppressive structures is not foreign to intersex persons. Intersex surgery has often been conducted without the consent of intersex people, and sometimes even without the consent of parents. This has been given tacit approval by the culture in general and sometimes also by the church. Legal recognition of intersex and validation of intersex are also in doubt in some countries. Jesus did not choose a time without oppression to become human. He identified with those who are oppressed to demonstrate God's power to overcome that oppression. However, Jesus did not come to overthrow the Romans, only to have another oppressive power take their place. The freedom that Jesus gives is not political freedom, but freedom to rise above circumstances. This is a far more powerful freedom than simply being free of negative circumstances, since it transcends circumstances. Jesus freed people through forgiveness, acceptance, and reconciliation.

A Perilous Childhood

In the life of Jesus, circumstances were not glorious, but rather the opposite. Like all humans Jesus began life as an infant. His infancy was a time of great vulnerability, even outright peril. According to Luke, Jesus' parents were forced to travel to Bethlehem because of a census, taken "in the days of

4. Grabbe, *An Introduction to First Century Judaism*, 1–19.

5. Borg and Wright, *The Meaning of Jesus*, 32.

6. Green, *The Gospel of Luke*, 64.

7. Keener, *The Gospel of Matthew*, 100.

Herod, king of Judea" (Luke 1:5a), a tyrannical client-king. They travelled to Joseph's home town of Bethlehem at the demand of Caesar Augustus, probably to pay the Roman poll tax. Living arrangements in Bethlehem were less than satisfactory. Jesus was placed in a manger when he was born (2:7). Possibly the family stayed with relatives or friends. At that time most poor people did not have separate accommodations for their animals, but lived under the same roof. One very old tradition claims that Jesus was born in a cave, and another that he was born in a courtyard. Whichever of these options is correct, Jesus began his life in poverty and obscurity.[8]

The situation recorded in Matthew's gospel was considerably more perilous. Magi came from the East seeking "he who has been born king of the Jews," and went to King Herod to discover the location his birth (2:1–2). The possibility of a new king disturbed Herod and others in positions of power in Jerusalem (2:3). So Herod, determined to eradicate the threat to his reign, entreated the Magi to tell him where to find the Messiah so that he could worship him (2:7–8). The ruse was made known to the Magi in a dream (2:12). Fortunately, Joseph was also warned in a dream to flee with his family to Egypt in order to escape Herod's murderous intentions (2:13–15).

> Then Herod, when he saw that he had been tricked by the wise men, became furious, and he sent and killed all the male children in Bethlehem and in all that region who were two years old or under, according to the time that he had ascertained from the wise men. (Matt 2:16)

Having been born in poverty and escaped being killed as a small child, Jesus then became a refugee in Egypt for some time.

> Now when [the Magi] had departed, behold, an angel of the Lord appeared to Joseph in a dream and said, "Rise, take the child and his mother, and flee to Egypt, and remain there until I tell you, for Herod is about to search for the child, to destroy him." And he rose and took the child and his mother by night and departed to Egypt and remained there until the death of Herod. This was to fulfill what the Lord had spoken by the prophet, "Out of Egypt I called my son." (Matt 2:13–15)

Egypt was a place of last resort for the poor (2 Macc 5:8–9). The hasty departure meant taking few possessions with them. Thus as a child, as if in

---

8. Green, *The Gospel of Luke*, 64, 126, 28–29; Morris, *Luke*, 92.

anticipation of his adulthood, Jesus had nowhere to lay his head (cf. Matt 8:20). When they returned to Israel the political situation was still far from ideal. Archelaus now reigned in place of his father, and he was just as unpleasant a ruler and less competent. Therefore, the family was uprooted again and went to live in Galilee in the town of Nazareth. Nazareth was a small, obscure, politically irrelevant town (2:19–23).[9] Therefore, Jesus spent his earliest years in poverty and obscurity, hiding from violent men, who desired to take his life even as a small child.

Jesus' infancy was a time of great vulnerability and peril as is the case with many intersex individuals. He was a defenseless infant with no control over his own circumstances. People with intersex traits often have fraught childhoods, and have endured much which is out of their control. During the last half-century, society, doctors, and sometimes parents, have tried to eradicate intersex through surgery and secrecy, and forced intersex people to hide who they are. Jesus knows the pain of having evil men try to eradicate his existence and then being forced to hide away out of sight away from his home. In experiencing these things, Jesus has identified with the perilous childhoods of intersex people.

The Pain of Childhood Mockery

When Jesus had survived the hazards of his infancy he then faced the psychological traumas of childhood. One likely trauma was the stigma attached to accusations of illegitimacy. Matthew tells us that Mary was betrothed to Joseph but was already pregnant before they were married. Joseph was clearly aware of her pregnancy and wanted to divorce her quietly. He agreed to marriage only after an angel of the Lord appeared to him (Matt 1:18–25). According to Luke, after Mary became pregnant by the Holy Spirit (Luke 1:35–38), she went to visit her relative Elizabeth. Mary then returned to her own home (1:56), not the home of Joseph, which you might expect if she were married. When Joseph took Mary to Bethlehem, she was not called his wife but his betrothed, yet she was expecting a child (2:5). During Jesus' visit to his home town (Mark 6:1–6) Jesus is called "the son of Mary" (6:3). Gerd Lüdemann argues that no one would call someone by the mother's name unless the son was illegitimate.[10]

9. Keener, *The Gospel of Matthew*, 107–13.

10. Lüdemann, *Virgin Birth?*, 51–53.

The people who knew Mary and Joseph when Jesus was a child doubtless did not believe Mary's claim to a virgin birth. She would have been the only one who could be sure of the truth. Others would have scoffed at the idea, since it is distinctly a one-off event. Probably people assumed she had become pregnant by Joseph, or worse still by someone else. In which case, Jesus grew up with the stigma of illegitimacy. If this is true, then no doubt he would have been teased as a child. Perhaps he was called a "bastard." In that culture this label had far more negative force than in present Western liberal society. The apocryphal book the Wisdom of Solomon records the stigma attached to being illegitimate in the time of Jesus' birth:

> But children of adulterers will not come to maturity, and the offspring of an unlawful union will perish. Even if they live long they will be held of no account, and finally their old age will be without honor. If they die young, they will have no hope and no consolation on the day of judgment. For the end of an unrighteous generation is grievous. (Wis 3:16–19)

So growing up in a small town, Jesus probably had to contend with being laughed at and mocked. No one can control the way in which they are born. But this does not stop children from being cruel. Teasing is a common experience for intersex people. One person with CAH reports, "I have spent a great deal of my life being teased about how I am developed. Lots of jokes; it has pretty much left me with a low acceptance of myself as a male."[11] One person with Klinefelter syndrome expresses a similar experience. "I was often teased for having small testicles, and I had gynecomastia (breast growth in a male)."[12] Jesus bore the mockery alongside all the intersex individuals who have lived with hurtful remarks because of the way that they were born.

An Intersex Jesus?

Since I have been discussing ways in which Jesus' infancy and childhood are like the childhood of intersex persons, it is important to consider one more matter. Is Jesus intersex? Biologist Edward Kessel has suggested a biological interpretation of the virgin birth of Jesus. He argues that since Jesus was born of a virgin he must have had XX chromosomes. Parthenogenesis

11. Callahan, *Between XX and XY*, 72.

12. Cameron, "Caught Between," 92.

(virgin birth) implies that the embryo has only female chromosomes. Consequently, Jesus was biologically female throughout his life. Since he appeared to be male, a process of sex reversal must have occurred in the womb. In this sense Jesus was both female and male, an androgynous Christ. Kessel's conclusion is that the androgynous Jesus renders arguments against the ordination of women null and void.[13] Some want to use Kessel's argument as a basis for speaking about an intersex Jesus.[14]

I am not a biologist and do not presume to argue with Kessel's logic. I will not deny that his argument seems to be carefully considered, even if his thesis that Jesus is both male and female is somewhat unlikely in itself. However, my concern lies with the idea that Jesus must be intersex in order to give credence or value to people who are intersex. I have argued in the previous chapter that there is no need for Jesus to be female to represent women or intersex to be the Savior of intersex persons. Such arguments are merely examples of the scandal of particularity, that is, the scandal that God has revealed himself in one particular person to the exclusion of all other possibilities. Instead of rehashing that argument, I will consider the actual significance of the virgin birth, and consequently explain why it is not important to argue that Jesus was intersex.

The virgin birth is not intended as an explanation of *how* the Son of God became human, but a description of *what* happened. It is thus impossible to understand the virgin birth biologically, since the birth of the Son of God is not only a matter of natural processes. Instead we must consider the wider truth of the mystery of Christ and his uniting of divinity and humanity into his person. This must be considered in concert with Jesus' resurrection from the dead, since both are signs of his continuity and discontinuity with fallen humanity. The virgin birth of Jesus points to his genuine humanity, begun in the womb of a mother. Yet at the same time, Jesus is quite different to other humans, since he existed beforehand as the Son of God. The Son of God did not unite himself to some human who already existed. His birth was an act of pure grace on the part of God. This unique birth began the new creation, which was completed when Christ was raised from the dead. The virgin birth and the resurrection point to the same reality, that God has acted in Christ to redeem the world.[15]

---

13. Kessel, "A Proposed Biological Interpretation of the Virgin Birth."

14. E.g., Mollenkott, *Omnigender*, 115–18.

15. Torrance, *Incarnation*, 94–104.

Therefore, however plausible Kessel's logic may be, his argument is actually irrelevant to the nature of salvation. It is not necessary that Jesus be both male and female, or intersex, in order for intersex persons to be saved or even that they are significant to God. The significance of the virgin birth is far greater than this. In the virgin birth the Creator entered the human race in order to transform the human condition. Jesus does not merely give acknowledgement to intersex persons through his (hypothetical) biology. In becoming human he wrought a complete transformation of humanity, overcame sin and death, and reconciled humanity to God. To concentrate on arguments that Jesus was potentially intersex is to miss the point of the incarnation entirely. He united himself with *our* humanity. He did this by becoming a particular human being, who can genuinely represent *every* human being, whether male, female, or intersex. Let us not trivialize the significance of the person and work of Christ.

The Ministry of Jesus

The infancy and childhood of Jesus involved much that is similar to the experiences of intersex people: powerlessness, vulnerability, danger, and mockery. These demonstrate Jesus' identification with intersex persons in their childhood years. The ministry of Jesus shows us how Jesus would have treated intersex people if he had met any during his time on earth. The Gospels do not mention any encounters between Jesus and intersex people. However, Jesus came to minister to the marginalized and the outcast. The Gospel accounts contrast the way Jesus treated people and the way marginalized people were treated by the general culture.

Good News to the Outcast

The ministry of Jesus takes up a significant part of the Gospels. Jesus did not simply come to die on a cross, but also to meet human need as he ushered in the kingdom of God. In Luke's gospel Jesus' first public act after his baptism was to define the parameters of his ministry. After fasting for forty days in the desert, he went to his home town of Nazareth and attended the synagogue on the Sabbath, where he read from the scroll of Isaiah (Luke 4:14–16).

> The Spirit of the Lord is upon me, because he has anointed me to proclaim good news to the poor. He has sent me to proclaim liberty to the captives and recovering of sight to the blind, to set at liberty those who are oppressed, to proclaim the year of the Lord's favor. (Luke 4:18–19)

This passage sums up the purpose of Jesus' ministry. His message and ministry were good news to the poor, the captives, the disabled, and the oppressed. In the ancient Mediterranean world, "poor" did not necessarily mean economically poor. Rather poverty was connected with education, family status, work, ritual purity, wealth, and gender. People without sufficient means were poor and so were those without status, those who were marginalized, and those outside the social and religious boundaries that defined honor. Jesus gave sight to those who were physically blind and to those who needed revelation. The liberty that Jesus brings encompasses both release from bondage and forgiveness of sins. The ministry of Jesus ushers in the end-time age of salvation, God's coming to his people.[16]

We might divide the people to whom Jesus ministered into two very broad categories: the sick and marginalized, and those who were outcast and labeled "sinners." These groups may well overlap, but for the sake of simplicity I will consider the groups one at a time. Of course Jesus ministered to a great many people, but these examples will be sufficient to demonstrate that Jesus was a man of compassion, who drew people from the outside to the inside. He made outcasts his friends. He reconciled sinners back to God. Jesus was not constrained by the social mores of his culture, which marginalized many people. Instead he broke through those customs, restoring people to wholeness and dignity, reconciling relationships, reinstituting the outcasts to community, and calling sinners to repentance through forgiveness of sin.

The Sick and Marginalized

> When he came down from the mountain, great crowds followed him. And behold, a leper came to him and knelt before him, saying, "Lord, if you will, you can make me clean." And Jesus stretched out his hand and touched him, saying, "I will; be clean." And immediately his leprosy was cleansed. And Jesus said to him, "See that you

16. Green, *The Gospel of Luke*, 211–12.

say nothing to anyone, but go, show yourself to the priest and offer
the gift that Moses commanded, for a proof to them." (Matt 8:1–4)

The leper was possibly the most marginalized person in Jewish society. Biblical leprosy is most likely not what we now call "Hanson's disease,"
but rather a variety of skin diseases, which cause bleeding, foul-smelling
sores on the body. Importantly, once a person was diagnosed by the priest
with leprosy, he or she was declared "unclean" and had to wear torn clothes,
keep hair unkempt, cover the upper lip, and cry out "unclean" when other
people approached. This necessitated living outside the community and
away from the temple. Because of this the person with leprosy was like the
walking dead (see Lev 13, particularly vv. 45–46). The social isolation must
have been acute.[17]

No doubt Jesus was aware of the biblical law forbidding people to
touch a leper (Lev 5:3). Some rabbis believed that leprosy was the most
defiling condition; even being in the same house as a leper was thought
to make a person unclean.[18] Yet when the leper begged to be made clean,
Jesus reached out and touched him and made him clean. Jesus had no need
to touch the man in order to heal him. He would later display his power to
heal at a distance (Matt 8:5–13). In Mark's gospel, Jesus seems to change
places with the leper. He is forced to go out into desolate places (because
the leper would not keep quiet about what he had done), while the leper
re-entered society (Mark 1:44–45). The story, therefore, demonstrates the
compassion that Jesus had for the leper and the risk that he was willing to
take in touching him. But Jesus did not shy away from the leper, because his
desire was to restore the outcast.

Other stories show us Jesus' interaction with those who were considered unclean. The first is the story of the woman with the uterine
hemorrhage.

> And behold, a woman who had suffered from a discharge of blood
> for twelve years came up behind him and touched the fringe of his
> garment, for she said to herself, "If I only touch his garment, I will
> be made well." Jesus turned, and seeing her he said, "Take heart,
> daughter; your faith has made you well." And instantly the woman
> was made well. (Matt 9:20–22)

---

17. Hartley, *Leviticus*, 190; Keener, *The Gospel of Matthew*, 259–60.

18. Keener, *The Gospel of Matthew*, 260.

The woman shared much with the leper of the earlier story. According to the law, she was unclean as long as the bleeding lasted (Lev 15:25–27). Although not as restricted in movements as the leper, the woman should have avoided crowds, since anything or anyone she touched would become unclean. But the woman was desperate. She battled the crowds (Mark 5:24) and touched Jesus' cloak, because she believed that in doing this she would be healed. And she was. Her desperation was fueled by both the long-term physical and social effects of the condition. Long-term uterine bleeding would have prevented her from marrying if it began when she was an adolescent, or provided a reason for divorce if it happened after marriage. She would have been unable to have sexual intercourse while bleeding (Lev 18:19), and therefore unable to produce children as was expected of a Jewish woman. The woman would have therefore been sick, poor, and isolated from others, a very depressing situation.[19]

Jesus knew that the woman had touched his cloak, and did not chastise her, but instead commended her—"your faith has made you well" (Matt 8:22). He was not ashamed to be associated with the unclean woman, even though he would then have been considered unclean by the crowds around him. Indeed, Jesus seemed unconcerned about contracting uncleanness. Immediately after allowing the unclean woman to touch his cloak, he then went and touched a dead girl (Matt 8:18–19, 23–25; Mark 5:21–24, 35–43). When he arrived to heal the girl, the mourners had already arrived (Mark 5:35). Touching a dead body, like touching a leper or touching a bleeding woman, resulted in ritual uncleanness (Num 19:11–12). But Jesus came to embrace the brokenness of others so that the broken could be made whole. In allowing others to make him unclean he willingly identified with and healed their shame.[20]

The way in which Jesus treated those labeled unclean is in stark contrast to the treatment of infants with ambiguous genitalia. Ambiguous genitalia seem to engender disgust in many doctors. There are two kinds of disgust—primary and projective. Primary disgust is produced by things that may cause disease. Projective disgust, on the other hand, considers a particular class of people contaminated or contaminating. Doctors project disgust onto intersex infants and thus feel less need to treat them humanely. General medical ethics are therefore effectively abandoned when it comes to dealing with intersex births, since those who disgust us are seen as less

19. Ibid., 302.
20. Ibid., 301, 304.

than human.[21] But this is not the way in which Jesus treated those labeled unclean. He treated them as human beings worthy of human contact and love. He did not shame the unclean woman or heal the leper from a distance. Instead he allowed himself to make physical contact with unclean people. This is the opposite of disgust. If Jesus had encountered an intersex person he would not have been disgusted, but instead have accorded full human dignity to that person.

Another category of unclean people in ancient Israel were Gentiles (non-Jews). Gentiles were considered unclean because they were not part of the Jewish people, and thereby excluded from the people of God. But Gentiles were not rejected out of hand by Jesus. Two Gentiles are commended by Jesus as people of great faith. The first is a Roman centurion.

> When he had entered Capernaum, a centurion came forward to him, appealing to him, "Lord, my servant is lying paralyzed at home, suffering terribly." And he said to him, "I will come and heal him." But the centurion replied, "Lord, I am not worthy to have you come under my roof, but only say the word, and my servant will be healed. For I too am a man under authority, with soldiers under me. And I say to one, 'Go,' and he goes, and to another, 'Come,' and he comes, and to my servant, 'Do this,' and he does it." (Matt 8:5–9)

Romans were considered pagans who worshipped the emperor. The centurion was part of an occupying foreign army, who collected taxes to fund pagan worship.[22] However, Jesus did not judge the centurion on the basis of his occupation or his foreign birth. Rather, "When Jesus heard this, he marveled and said to those who followed him, 'Truly, I tell you, with no one in Israel have I found such faith'" (Matt 8:10). The centurion was held up to others as an example of someone who will enter the kingdom of God even while Jews were losing their place in it (8:11–12). The Gentile was commended as someone with more faith than those in Israel. The hated outsider was ushered into the kingdom by Jesus and given a new status because of his great faith.

The Canaanite woman from the vicinity of Tyre and Sidon (Matt 15:21–28) must have been equally on the outer margins. The Canaanites were the enemies of Israel and many times Israel was led away from their God because of the idolatry of Canaan. It would be surprising if this woman

21. Feder, *Making Sense of Intersex*, 75–77.

22. Keener, *The Gospel of Matthew*, 264–65.

did not feel excluded from those who counted. Why would a Jew even speak to her? But she came to Jesus and begged him to heal her demon-possessed daughter (15:22). At first Jesus appeared to test her by ignoring her request, even denying that she had a right to make it (15:23–26). Yet the woman persisted, "Yes, Lord, yet even the dogs eat the crumbs that fall from their masters' table" (15:27). Her daughter was healed and her humility was commended by Jesus: "You have great faith" (15:28).

In the time of Jesus, Jews saw themselves as privileged people, chosen by God. Many believed that they were favored by God solely on the basis of their biology (Matt 3:9; John 8:39). But Jesus embraced Gentiles as well as Jews: he ministered to the Canaanite woman and the Roman centurion (8:5–13 above); pagan astrologers acknowledged his birth (2:1–12); and indeed the genealogy of Jesus himself contains several Gentiles—Tamar the Canaanite (1:3), Rahab of Jericho (1:5), and Ruth the Moabite (1:5). Having the correct biology is not the means of gaining the favor of God. Instead Jesus is himself the only way to the Father (John 14:6), and God's favor can only be found through faith in him. Although some intersex believers have been told that they cannot be Christians because of their biology, Jesus did not privilege biology over faith, but rather privileged those who had faith in him and gave them a place among the genuine people of God.

The next two stories are concerned with sick people and sin. The first demonstrates that although many were content to blame the marginalized for being marginalized, Jesus did not think this way.

> As he passed by, he saw a man blind from birth. And his disciples asked him, "Rabbi, who sinned, this man or his parents, that he was born blind?" Jesus answered, "It was not that this man sinned, or his parents, but that the works of God might be displayed in him." (John 9:1–3)

While it seems to be a rather strange question to ask whether the blind man had sinned in the womb, this was not an unusual question in the ancient world. The Ten Commandments mention the sins of the parents being visited on the children (Exod 20:15; Deut 5:9). The Targum (an ancient Jewish commentary on the Old Testament) on Deuteronomy 21:20 suggests that parents get what they deserve in their children. Rabbis also debated about children sinning in the womb. The passage cited in this argument is Genesis 25:22, where Rebecca's twins struggled in the womb. Jesus rejected both options—the sin of the parents and the sin of the child in the womb—in this instance. This man had done nothing to deserve being blind from birth. Instead of blame, the blindness had a purpose, to display the works of God through his healing by Jesus (9:3). Jesus displayed his glory in healing the man and encouraged his disciples to use their limited time wisely to meet people's needs (9:4), rather than apportioning blame.[23]

The man born blind had not sinned in the womb. The person born intersex has not sinned in the womb. Intersex is simply a fact of life for that person and not a punishment for individual sin. In Chapter 2, I claimed that intersex is the result of the fall. By this I did not mean that intersex is the result of individual sins. Rather the fall has resulted in many different kinds of distortions in the world, physical and otherwise. Intersex is one of those. The ministry of Jesus demonstrates that he was far more interested in helping people than condemning them on the basis of things over which they had no control. The attitude of Jesus towards the blind man contrasts markedly with the religious leaders, who said to the man, "You were born in utter sin, and would you teach us?" and then threw him out of the synagogue (John 9:34). The attitude of the church needs to be like that of Jesus, loving people and meeting their needs, rather than making judgments about people based on the way they are born.

The second story apportions no blame to the sick man for his illness, but shows that forgiveness is more important than physical healing.

> And getting into a boat he crossed over and came to his own city. And behold, some people brought to him a paralytic, lying on a bed. And when Jesus saw their faith, he said to the paralytic, "Take heart, my son; your sins are forgiven." And behold, some of the scribes said to themselves, "This man is blaspheming." But Jesus, knowing their thoughts, said, "Why do you think evil in your hearts? For which is easier, to say, 'Your sins are forgiven,' or to say, 'Rise and walk'? But that you may know that the Son of Man has authority on earth to forgive sins"—he then said to the paralytic—"Rise, pick up your bed and go home." And he rose and went home. When the crowds saw it, they were afraid, and they glorified God, who had given such authority to men. (Matt 9:1–8)

The earning capacity of the paralytic was limited at best. As a man with a disability, his status in society was low. It was quite obvious to all that he had a real physical need of healing. It is interesting, therefore, that Jesus did not immediately heal the man as we might expect him to do. Instead, the first thing which Jesus said to the man was, "Take heart, my

23. Beasley-Murray, *John*, 152, 154–55.

son, your sins are forgiven" (Matt 9:2). The story of the man born blind in John 9 (above) indicates that Jesus did not always see sickness as the result of individual sin. There is no indication in Matthew 9 that Jesus attributed the disability to the man's sin. But he did perceive that forgiveness was of primary importance to the man.

Forgiveness is a primary need of all human beings, sick or well, male, female, or intersex. Before we need physical healing we need forgiveness and reconciliation with God. In the case of this man the physical healing was also granted, but it is not the most fundamental need. In the case of intersex people, physical healing may not be granted in the present life. Certainly many intersex people do not perceive medical intervention in their lives as healing, but rather invasive and even destructive. As Christians we must see forgiveness as much more important than bodily wholeness or unambiguous sex. The first thing that intersex individuals need is the forgiveness of Jesus. Whether people desire any physical changes in their bodies or not, God's forgiveness is primary.

Sinners in Need of Salvation

The mention of our need of forgiveness leads to the second significant group to which Jesus ministered—sinners in need of salvation. I am exploring sinners in need of salvation, not because intersex people are more in need of salvation than others, but because the interactions which Jesus had with sinners evidence his grace and acceptance of people. This grace and acceptance drew people to Jesus and resulted in genuinely transformed lives.

In the social world of ancient Israel some occupations were scorned, tax collector being one of these. Following on in Matthew's gospel from the story about the forgiven and healed paralytic is a story about a tax collector named Matthew.

> As Jesus passed on from there, he saw a man called Matthew sitting at the tax booth, and he said to him, "Follow me." And he rose and followed him. And as Jesus reclined at table in the house, behold, many tax collectors and sinners came and were reclining with Jesus and his disciples. And when the Pharisees saw this, they said to his disciples, "Why does your teacher eat with tax collectors and sinners?" But when he heard it, he said, "Those who are well have no need of a physician, but those who are sick. Go and learn

what this means, 'I desire mercy, and not sacrifice.' For I came not
to call the righteous, but sinners." (Matt 9:9–13)

Tax collectors were hated by the ordinary members of the populace,
because they collected tax for the occupying Romans. Tax collectors often
collected more tax than was necessary and some even used violence against
people who did not pay up. Even without the extra money extracted by
tax collectors, the Roman taxes were a heavy burden to people.[24] No one
wanted to finance the occupying armies or be friends with the tax collectors
who sided with the enemy through their occupation.

Jesus was unconcerned about what others thought of the tax collector. He called the man to follow him. Then, in an even more outrageous
act, Jesus went to eat with the tax collector. Eating a meal with someone
was a symbol of approval of that person. So it is not surprising that the
Pharisees were outraged by this. Not interested in placating the Pharisees,
he responded to them, "Those who are well have no need of a physician,
but those who are sick. . . . I come not to call the righteous, but sinners." No
doubt the Pharisees thought of themselves as the righteous (see Luke 18:11)
and therefore did not perceive any need for repentance (cf. Matt 21:23–27)
or to follow Jesus. Tax collectors and sinners, on the other hand, were aware
of their status as outcasts. It was easier for them to recognize their need of
Jesus.

The tax collector was an outcast for a reason, because he was considered a traitor. Intersex people can be outcasts for no reason other than
the fact that they are different. The secrecy that has surrounded intersex
conditions has reinforced the idea that intersex makes a person a pariah.
The secrecy and the surgeries that try to cover up its existence speak loudly,
implying that intersex is something shameful, cannot ever be spoken about,
and cannot come into the open. But if Jesus loved the tax collector, he most
certainly loves the intersex, who are on the outside for no good reason. Jesus ate with the outcasts and he would fellowship with the intersex, because
he loves the marginalized and rejected, and calls them to follow him.

Tax collectors were hated for their occupation. Possibly even more
contempt was felt by the religious establishment for a woman caught in
adultery. And just as much mercy was shown by Jesus to her.

The scribes and the Pharisees brought a woman who had been
caught in adultery, and placing her in the midst they said to him,

24. Keener, *The Gospel of Matthew*, 292–93.

"Teacher, this woman has been caught in the act of adultery. Now in the Law Moses commanded us to stone such women. So what do you say?" This they said to test him, that they might have some charge to bring against him. Jesus bent down and wrote with his finger on the ground. And as they continued to ask him, he stood up and said to them, "Let him who is without sin among you be the first to throw a stone at her." And once more he bent down and wrote on the ground. But when they heard it, they went away one by one, beginning with the older ones, and Jesus was left alone with the woman standing before him. Jesus stood up and said to her, "Woman, where are they? Has no one condemned you?" She said, "No one, Lord." And Jesus said, "Neither do I condemn you; go, and from now on sin no more." (John 8:3–11)

The Pharisees had no compassion for the woman at all; she was merely a pawn in their plot to discredit Jesus. Since they had been watching Jesus, the Pharisees would have been aware of how Jesus treated sinners and derelicts with compassion and mercy. They used the woman to trick Jesus into saying the wrong thing. Either he would uphold the law that commanded the woman to be stoned to death (Lev 20:10) and thereby contradict the way he had acted previously, or he would deny that the woman needed to be stoned and thus go against the law. Either way they believed that he would trip up.[25] Jesus' actions allowed them no satisfaction. Instead they were all convicted as sinners. Jesus did not condemn the woman, but called her to leave her life of sin and walk free.

There are two aspects of this story that may be applied to intersex people. The first is that Jesus' ministry was a ministry of grace. The law imposed a penalty on the woman, but Jesus did not apply that penalty. Instead, he offered her overwhelming grace. Overwhelming grace must be applied to intersex people. This is applicable firstly in the realm of gender. The biological reality, unwanted surgery, and parental choices can lead to personal confusion over gender. In the sphere of grace the person can come to a place of peace about which gender to live as. In a place of overwhelming grace there can be support through this decision-making process. When the church grasps the grace which flows from Jesus, the one who is "full of grace and truth" (John 1:14), then the church can be that place of grace.

The second important aspect of this story is the way in which Jesus understood the Scriptures. The religious leaders were very concerned about every possible detail of the law. They "built a fence around the law," making

25. Beasley-Murray, *John*, 143–44, 146.

the commands stricter than those given by God. Jesus interpreted the law according to its original intent. For example, his understanding of the Sabbath was completely different from those who refused his hungry disciples food while walking on the Sabbath (Matt 12:1–8), and who denied that Jesus could lawfully heal the sick on the Sabbath (12:9–14).[26] In the case of the woman caught in adultery, a strictly literal reading of the law would have meant that the woman should have been stoned and Jesus should have joined in. Yet Jesus interpreted the Scripture through the eyes of grace.

A strict interpretation of the Scriptures, particularly Genesis 1:26–27, may result in the conclusion that since intersex people are neither clearly male nor clearly female then they are not in the image of God. But Jesus interpreted the Scriptures in light of God's love, grace, and mercy. We must therefore be careful not to use the Bible as a tool for controlling people through religion or for condemning people. Rather, we interpret the Bible through its center, namely Christ, because then we will be drawn to a relationship with God. Grace reminds us that it is not the things over which we have no control, like ambiguous biology, which define us as human. We are human because of the gracious gift of life that comes from Jesus.

Conclusion

The Son of God chose to become human, and in doing so he identified with the human experience. Much of Jesus' life was in many ways like the lives of intersex people. He experienced vulnerability, peril, and prejudice. He was not intersex, but he understands the pain that many people with ambiguous biology feel. Jesus did not minister to any intersex people and yet we know how he would treat intersex persons. The marginalized, the rejected, the unclean, the sinners, and the outcasts were all loved, dignified, restored, included, and humanized by Jesus. That is what Jesus would do for any intersex person he met.

The life and ministry of Jesus should be an example for the church to follow. If the church wants to be like Christ, then our priority must be to accept difficulties and insults alongside those who are marginalized because they are different. The church is called to serve the weak and outcast, not to condemn them. The ministry of Jesus was one of acceptance, reconciliation, and healing. Therefore, the church is called to heal intersex people, not physically, but by welcoming them and accepting them. Inclusion is

26. Keener, *The Gospel of Matthew*, 348, 350.

the opposite of shame and disgust. Inclusion and love will offer grace and support as we walk alongside people who are trying to work through significant issues in their lives.

Of course a discussion of the life of Jesus cannot be complete without a discussion of his death. Therefore, in the next chapter I will continue to discuss the work of Jesus, specifically his work of atonement, beginning with the persecution and betrayal of Jesus. Then I will explore the events of the crucifixion and how these impact upon issues that specifically relate to the experience of intersex people: shame, forgiveness, identity, worship, and gender.

5

The Cross and Intersex

Introduction

THE LIFE OF JESUS was one of identification with the human experience, including its pain and difficulty, both physical and emotional. His ministry was compassionate, healing, inclusive, loving, and dignifying. But there is more to the work of Jesus than his life and ministry. Christians have always attributed great significance to the death of Jesus, because his death completes his work of forgiveness and reconciliation with God. Without his death, humanity could not have free access to God, but would be eternally in bondage to sin. It is not possible for this short chapter to explore the death of Christ fully. Although important aspects of the atonement may go unstated, these are nonetheless implicitly assumed. I do not intend to produce a treatise on the atonement in all its wonder, but rather to focus on particular aspects of the death of Jesus that specifically impact intersex issues.

Jesus experienced both persecution and betrayal during his ministry, before being given over to death. He was hated by religious leaders and betrayed by a close friend. This gives Jesus an insight into the betrayal and persecution that has been experienced by some intersex people. The death of Jesus also has importance for intersex people in particular. Jesus bore the shame of intersex in his passion. He offered forgiveness and calls intersex people to forgive those who hurt them. New identity is offered because of the cross. Jesus' death overrides biological restrictions on worship. And lastly, the cross is the penultimate step in fulfilling the purpose of sex and gender, and brings healing to broken sexuality.

Persecution and Betrayal

The positive ministry of Jesus to the outcast, the marginalized, and the sinner, was in contrast to his often negative relationship to people of power. From his infancy and through his ministry, Jesus was persecuted. This persecution escalated as time went on. Eventually the desire of the religious establishment to get rid of Jesus was made possible when one of Jesus' close friends betrayed him. Because Jesus has undergone both persecution, and betrayal at the hands of a friend, he is well placed to empathize with those persons with intersex traits who have been persecuted and feel betrayed.

As Luke tells the story, Jesus began his ministry in his home (Luke 4:16). Although the people were amenable to begin with, they soon began to ask, "Isn't this Joseph's son?" (4:22). How could this man who grew up in this place be a prophet? Jesus said to them, "Truly, I say to you, no prophet is acceptable in his hometown" (4:24). It was not long before the people of Nazareth "rose up and drove him out of the town and brought him to the brow of the hill on which their town was built, so that they could throw him down the cliff" (4:29). But he walked through the crowd and went away (4:30).

The situation simply grew grimmer as the ministry of Jesus progressed. He antagonized the religious leaders by healing on the Sabbath. The Pharisees wanted to find a reason to accuse Jesus so they watched him closely (Mark 3:2) and were quick to plot against him with the Herodians (Mark 3:6; cf. Luke 6:11). When he had exposed the hypocrisy of those same leaders (Luke 11:39–52), the Pharisees and Scribes "began to press him hard and to provoke him to speak about many things, lying in wait for him, to catch him in something he might say" (11:53–54). But Jesus continued on to Jerusalem "for it cannot be that a prophet should perish away from Jerusalem" (13:33b). When Jesus had reached Jerusalem he made the leaders more determined to kill him when he drove the sellers out of the temple. They plotted but could not think of a way to get rid of Jesus while the people were eager to hear him (19:45–48).

The opportunity for the religious leaders to have Jesus killed finally arrived when Judas agreed to betray him. Jesus knew in advance that he would be betrayed by one of his disciples.

I am not speaking of all of you; I know whom I have chosen. But the Scripture will be fulfilled, "He who ate my bread has lifted his heel against me." . . . After saying these things, Jesus was troubled in his spirit, and testified, "Truly, truly, I say to you, one of you will betray me." . . . "It is he to whom I will give this morsel of bread when I have dipped it." So when he had dipped the morsel, he gave it to Judas, the son of Simon Iscariot. Then after he had taken the morsel, Satan entered into him. Jesus said to him, "What you are going to do, do quickly." (John 13:18, 21, 26–27)

It is shocking that the person who betrayed Jesus was someone so close to him. The enemies of Jesus had been plotting to kill him for some time, even from infancy. But the betrayal of a close friend must have been far more painful. Some understand the expression "lifted up the heel" as a sign of contempt, suggesting that Judas harbored a deep hatred of Jesus. But Jesus showed no sign of hating Judas. Rather he dipped bread in the dish and shared it with Judas as a mark of friendship.[1]

Jesus was no stranger to persecution. He had done nothing to warrant it, except offend the religious and political sensibilities of those around him. This led to his persecution, betrayal, and death—and a horrible, shameful death at that. Because of this, Jesus understands the persecution of people who are intersex, something that unfortunately does happen. The United Nations High Commissioner for Refugees asserts:

Intersex individuals may endure persecution because they do not conform to gender expectations, or are viewed as having a physical disability related to their atypical sexual anatomy. Family members of intersex persons are sometimes also abused. Intersex individuals may be subjected to unwanted surgery to "correct" their anatomy or have ongoing medical needs related to their condition.[2]

Persecution of intersex athletes is also evidenced in the media. A particularly famous case is that of South African runner Caster Semenya, who was found to be intersex during a test conducted by the IAAF (International Association of Athletics Federations). The experience of being accused of dishonesty in sport and having her biology exposed to the public gaze left her traumatized and she went into hiding. Another intersex athlete,

1. Beasley-Murray, John, 236–38.

2. United Nations High Commissioner for Refugees, Working with Lesbian, Gay, Bisexual, Transgender and Intersex Persons in Forced Displacement, 6.

put through the same public humiliation, attempted suicide.[3] Because Jesus has experienced unmerited persecution, he can identify with those intersex persons whose lives have been severely disturbed by such experiences.

Jesus definitely understands what it is to be betrayed by someone very close to him. Some intersex persons feel betrayed, not just by doctors, but also by parents. One example is found in the story of Ruby and her daughters. Two of Ruby's daughters have CAH. Both underwent genital surgery. Even though Ruby did not conduct the surgery, and even though Ruby was constrained in her choices when trying to get medical treatment for her daughters, her younger daughter was estranged from Ruby for a long time,[4] because she felt betrayed. The hurt over being betrayed by someone close is great, but, because he experienced his own betrayal, Jesus empathizes with intersex persons as they go through this experience.

Crucifixion

The inevitable result of the persecution and betrayal of Jesus was his crucifixion at the hands of the Romans. Because of the death of Jesus, humanity is reconciled to God (Col 1:20) and condemnation is undone (Rom 8:1). The significance of the atonement cannot be understated. However, my primary goal here is to concentrate on those aspects of the death of Christ that are particularly applicable to intersex. I do not intend to detract from the wider significance of the cross, only to focus on specific aspects. The particular matters that pertain most directly to intersex are: undoing shame, granting forgiveness, new identity, new access in worship, and bringing the purpose of sex and gender to fulfillment.

Shame

The death of Jesus primarily dealt with our guilt and our shame. That every person needs to be freed from guilt and shame must go without saying. People who are intersex are no less guilty and no more guilty than the unambiguously sexed. However, shame is a problem that affects many intersex people. Possibly it is *the* intersex issue. In part, the shame is imposed on them by others who find their ambiguous sexual biology unacceptable. For many, that shame has been internalized and has led to depression and

3. Iqbal, "The Persecution of Caster Semenya".
4. Feder, *Making Sense of Intersex*, 52–54, 62.

self-harm. What Jesus experienced in his passion (both his crucifixion and the events just prior to it) was shame of the most intense kind. Crucifixion was intended to degrade, to humiliate, and to remove all vestige of humanness from the victim. Death on a cross was the most shameful of deaths, never mentioned in civilized society. Jesus' bearing of intersex shame extended even to knowing the pain of being naked and exposed, having his genitals on display and open to mockery.

When Adam and Eve sinned and first experie nced shame (Gen 3:6–7) they were punished for their sin (3:16–24), but God was merciful to them and covered their nakedness with animal skins (3:21). The punishment for humanity was mitigated by God. But in the passion of Jesus, there was absolutely no mitigation. Jesus took the full brunt of the shame and guilt that should fall on sinners.

In his passion Jesus was taken to the depths of human misery, despair, and indignity. He was treated as something inhuman, as an object with no right to dignity. In Matthew's account, Pilate yielded to the crowd and had Jesus flogged and then crucified (Matt 27:24–26). Flogging was intended to degrade as well as to cause pain. According to one description:

> The delinquent was stripped, bound to a post or a pillar, or sometimes simply thrown on the ground, and beaten by a number of torturers until the latter grew tired and the flesh of the delinquent hung in bleeding shreds. In the provinces this was the task of the soldiers. Three different kinds of implements were customary. Rods were used on freemen; military punishments were inflicted with sticks, but for slaves scourges or whips were used, the leather thongs of these being often fitted with a spike or with several pieces of bone or lead joined to form a chain. The scourging of Jesus was carried out with these last-named instruments. . . . It is not surprising to hear that delinquents frequently collapsed and died under this procedure, which only in exceptional cases was prescribed as a death sentence.[5]

The flogging was followed by more degradation. The soldiers dressed Jesus up like a king with a robe and a crown, but the crown was made of thorns, possibly twelve inches long. They mocked his claim to kingship with false worship, spat on him, and hit him over the head with a staff (Matt 27:27–31).[6] Spitting on Jesus would have communicated the soldiers' utter

---

5. Blinzler, *The Trial of Jesus*, quoted in Beasley-Murray, *John*, 335–36.
6. Beasley-Murray, *John*, 336.

disgust for him. From the perspective of Jesus, Gentile spittle was unclean, making the experience even more degrading.[7] After the flogging and mocking worship, "they led him away to crucify him" (27:31).

The Gospels do not go into the details of the crucifixion itself. The mention of Jesus on the cross would have been distressing to the original readers, who knew the horror of crucifixion.[8] It was considered the height of inappropriateness for any civilized person to mention the word "cross." First-century–BC philosopher Cicero wrote:

> But the executioner, the veiling of the head and the very word "cross" should be far removed not only from the person of a Roman citizen but from his thoughts, his eyes and his ears. For it is not only the actual occurrence of these things or the endurance of them, but liability to them, the expectation, indeed the very mention of them, that is unworthy of a Roman citizen and a free man.[9]

Crucifixion was therefore considered by all to be an utterly shameful death.

Because crucifixion was intended to be a humiliating and shameful mode of execution, not merely a painful one, the criminal was crucified naked. This would have been intensely shameful for a Jewish man. The Gospels do not mention mockery of Jesus' circumcision by the Roman soldiers, but there is no reason to assume that hardened and sadistic Roman soldiers would have been polite to a naked, dying Jew. Greeks and Romans believed circumcision to be a sign that Jews were barbaric, backward, and superstitious. Both Greeks and Romans banned circumcision at different times, with quite severe penalties attached. The circumcised penis was considered both aesthetically displeasing and morally indecent. Even some Jews found circumcision unsightly and underwent medical treatments to cover their circumcision in order to assimilate into Greco-Roman society.[10] For these reasons it is quite likely that Jesus experienced derision about his naked genitals as he hung dying and in pain.

Jesus underwent this experience of extreme shame on behalf of intersex people, whose lives have in many cases included great shaming. Whatever shame is attached to intersex has been borne by Jesus in his passion. He bore the shame of social stigma attached to his death. He bore the shame of being treated as something less than human. He bore the shame of being

7. Keener, *The Gospel of Matthew*, 675–76.

8. Ibid., 677.

9. Hengel, *Crucifixion*, 42.

10. Hodges, "The Ideal Prepuce in Ancient Greece and Rome," 387–91, 393, 399.

unable to cover his own body. He bore the shame of having his genitals exposed and mocked. He did this in order to heal the shame of intersex, the shame of being told that you are a freak, the shame of being subjected to repeated medical exams, the shame of being told your genitals are unacceptable, the shame of feeling rejected by parents. All this shame was taken by Jesus. In taking the shame of intersex, he has declared by his actions that intersex is not something shameful. Being intersex is a valid way of being human. Intersex is not unlovable and unworthy, since the death of Christ makes clear that intersex is no longer shameful. Since shame has been borne by Jesus its opposite must now be true, that is, intersex is accepted by God as without shame.

Forgiveness

Forgiveness of sin is the principal fruit of Jesus' death. We are all sinners who desperately need to be forgiven. And God forgives us based on the shed blood of Jesus. Forgiveness is offered to unambiguously sexed people and intersex people on the same basis. Jesus even forgave the people who caused him intense shame. Such mercy should compel us to offer forgiveness to others, regardless of what others have done.

Jesus prayed on the cross, "Father, forgive them, for they know not what they do" (Luke 23:34a). In asking the Father to forgive those who crucified him, he was acting out his main reason for coming into the world. Both intersex and unambiguously sexed people are sinners in need of forgiveness. Importantly, forgiveness of sin is not tied to any biological trait— neither race nor gender nor age is relevant to forgiveness. There is only one way to receive the forgiveness offered by God in the new covenant, that is, through faith in Jesus Christ.

> Blessed be the God and Father of our Lord Jesus Christ, who has blessed us in Christ with every spiritual blessing in the heavenly places, . . . *In him* we have redemption through his blood, the forgiveness of our trespasses, according to the riches of his grace. (Eph 1:3, 7)

This may seem like stating the obvious, but it is nonetheless worth stating in the face of ignorance in the church about intersex. It appears that some people have not grasped this fundamental point. One intersex Christian observes:

I am a Christian, therefore the views of other Christians are bound to mean something to me. . . . You know, when you're told by Christians who you respect "You are not a Christian," . . . that's quite damaging. And yes, it's because I am a Christian that that has had such an effect on me. . . . I've been told I'm not a Christian because of a condition I've got, by another Christian I respect. Yes, that's bound to affect me, isn't it?[11]

Therefore, I reiterate that the way in which a person who is intersex receives forgiveness from God is the same way in which a person who is unambiguously sexed receives forgiveness of sins, through Christ.

There is a further implication of the cross regarding forgiveness. The length to which God has gone—in giving up his only Son—in order to bring about unlimited forgiveness of sin has some implications for our own actions. Jesus told a parable about a man who was forgiven a great debt.

Therefore the kingdom of heaven may be compared to a king who wished to settle accounts with his servants. When he began to settle, one was brought to him who owed him ten thousand talents. And since he could not pay, his master ordered him to be sold, with his wife and children and all that he had, and payment to be made. So the servant fell on his knees, imploring him, "Have patience with me, and I will pay you everything." And out of pity for him, the master of that servant released him and forgave him the debt. But when that same servant went out, he found one of his fellow servants who owed him a hundred denarii, and seizing him, he began to choke him, saying, "Pay what you owe." So his fellow servant fell down and pleaded with him, "Have patience with me, and I will pay you." He refused and went and put him in prison until he should pay the debt. When his fellow servants saw what had taken place, they were greatly distressed, and they went and reported to their master all that had taken place. Then his master summoned him and said to him, "You wicked servant! I forgave you all that debt because you pleaded with me. And should not you have had mercy on your fellow servant, as I had mercy on you?" And in anger his master delivered him to the jailers, until he should pay all his debt. So also my heavenly Father will do to every one of you, if you do not forgive your brother from your heart. (Matt 18:23–35)

11. Seren (Baptist/Methodist) in "What Do Intersex Christians Say about Intersex and Faith?"

The debt owed in the parable was immense. The debtor owed more money than existed in the entire nation of Israel at that time, so he had no hope of repaying it.[12] The forgiveness offered to the servant was beyond incredibly generous. The debt was wiped out. But the man whose huge debt was forgiven failed to forgive the minute debt owed to him. The parable is of course not about money, but about forgiveness of sin. It emphasizes the immensity of forgiveness on the part of God, who has forgiven us of a debt so great that no one could ever repay in many lifetimes. Whatever another person may have done to us cannot be compared to the enormity of our own sin towards God. Therefore, because of the incredible mercy of God, we must be willing to forgive the sins of others against us.

Many intersex people have been sinned against a great deal. They have been violated through surgery, sinned against through secrecy, and betrayed by parents and medical personnel. Close relationships have been strained by these actions and many parents are estranged from children. There is great emotional hurt. The words that Jesus uttered from the cross are very pertinent to intersex people. Earlier I quoted only half of the verse. Luke 23:34 in full reads: "And Jesus said, 'Father, forgive them, for they know not what they do.' *And they cast lots to divide his garments*" (my emphasis). When Jesus asked the Father to forgive the soldiers, he forgave those who took away his dignity, who exposed his genitals to public scrutiny, and who deliberately heaped shame upon him. The parallel between the prayer of the cross and what Jesus asks intersex people to do should be clear. He forgave those who caused his shame and he asks intersex people to forgive those who have caused their shame.

Forgiveness is a difficult thing to offer. But Jesus calls his people to forgive as he forgave them (Eph 4:32; Col 3:13). He does not leave us to do this alone. Jesus has given his Spirit to believers so that we may obey his commands (Rom 8:1–13). Although forgiveness is difficult, it is possible in the power of the Holy Spirit. Forgiveness is not necessarily reconciliation. Reconciliation requires repentance on the part of the person who sinned against you. Yet it is possible to live without bitterness by forgiving the hurts that others have inflicted upon you, even if reconciliation does not result.

---

12. Keener, *The Gospel of Matthew*, 459.

Identity

Another very pressing issue for intersex persons is identity. But the cross is the place where human identity is transformed. At the cross Jesus went from being completely sure of who he is to being without an identity at all. In doing this, he made the way for sinners to find a new identity as sons of God through faith in him. Intersex need not be a hindrance to a solid identity if the intersex person embraces being a son of God through Christ.

For Jesus, possibly the predominant personal understanding of his identity was as the Son of the Father. At his baptism, his identity as Son of God was solidified.

> And when he came up out of the water, immediately he saw the heavens being torn open and the Spirit descending on him like a dove. And a voice came from heaven, "*You are my beloved Son; with you I am well pleased.*" (Mark 1:10, my emphasis)

From the beginning of his ministry, then, Jesus knew that he was the beloved Son, the one who his Father favored. He repeatedly used the term "Father" when he spoke of (Matt 12:50; 15:13; 16:17; 18:10, 19, 35; 20:23) or prayed to God (11:25–26). His relationship with the Father was utterly unique (11:27). Even as Jesus approached the time of his death, he held firmly to his identity as Son of God. He still referred to God as Father in the garden of Gethsemane (26:39, 42). Indeed, in Mark's account of Gethsemane, Jesus uttered the most intimate prayer to Abba, Father (Mark 14:36).

But there was a point while on the cross that Jesus lost his sure consciousness of divine Sonship.

> Now from the sixth hour there was darkness over all the land until the ninth hour. And about the ninth hour Jesus cried out with a loud voice, saying, "*Eli, Eli, lema sabachthani?*" that is, "My God, my God, why have you forsaken me?" (Matt 27:45–46)

During this time, Jesus had no experience of the presence of God. In his experience he was no longer the beloved Son in whom God, his Father, was well pleased. Instead he was godforsaken. The identity that had been his from childhood was in tatters. If Jesus did not know God as Father and himself as Son, then what identity could he possibly possess? At that moment of forsakenness, Jesus was without identity.

Jesus lost his identity because he bore the sins of others. One consequence of sin is an abiding confusion over identity. Humans should be secure in their identity as creatures made in the image of God. But sinners have rejected their God-given identity. In order to rescue humanity from the power of sin, Jesus was given up by the Father (Rom 8:32) and indeed gave himself up (Gal 2:20). In enduring the abandonment and godforsakenness of the cross, Jesus gave up his identity as Son of God for our sakes. He *lost* his consciousness of his identity as the Son so that each person who believes in him might *gain* an identity as a son of God through him (Gal 4:4–6).

For many intersex people, identity confusion is an everyday occurrence, exacerbated by ambiguous biology. Parents also experience confusion over the child's identity from the time that sexual ambiguity becomes evident. One parent described her confusion this way:

> As I remember it, we were told she was a boy, we were told she was a girl, and then we were told they didn't know. And that's kind of where I remember being left, as "we don't know." So having this baby that was a *nothing* so to speak, and then going to the ward, and being left in this state of not knowing really.[13]

The biological issues may not be resolved, but sonship is not about biology. Although many Jews in Jesus' day thought they were sons of God simply because they were biologically descended from Abraham (Matt 3:9; John 8:39), being a son of God can only happen through trusting in the unique Son, Jesus.

Because on the cross Jesus lost his identity as Son of God, believers can call God Father (Gal 4:6). Sons of God through Christ share a common Father in heaven and are part of the one family of God. So being a son of God means relationship with God as Father and relationship with other members of the family of God, the church. Ambiguously sexed and unambiguously sexed alike must live together as the family of God. The church should be a place of security, where each person is loved as a fellow brother or sister of Jesus. In a family there should be no need to hide who we are. Family is a place of belonging. The challenge to the church is to embrace intersex believers like the brothers and sisters of Jesus they are.

---

13. Anna quoted in Gough et al., "'They Did Not Have a Word,'" 499. Italics original.

New Access in Worship

The cross of Christ has opened up access to worship for all people. Access to God no longer has anything to do with biology. Although restrictions on worship existed in the Old Testament for eunuchs, Gentiles, and some people with disabilities, these are undone by Jesus. Therefore, intersex conditions cannot prevent anyone from coming to God in worship.

At the point when Jesus died, something very significant happened that opened the way to God for all people. "And Jesus uttered a loud cry and breathed his last. And the curtain of the temple was torn in two, from top to bottom" (Mark 15:37–38). The import of this event cannot be understood without referring back to the Old Testament temple. The innermost part of the temple was hidden behind a curtain. Only the high priest could enter behind the curtain and he could only enter on the Day of Atonement (Lev 16; Heb 9:7). The high priest had to be biologically male, an Israelite, from the tribe of Levi, a descendent of Aaron (Num 3:10), between the ages of thirty and fifty years (Num 4:3, 23, 30, 35), and have no physical deformities (Lev 21:16–23). The curtain of the temple, then, represented the most significant of all barriers to worship of God. Therefore, the removal of the curtain means that there is now universal access in worship through Christ.

This new access was foreshadowed by Jesus during his ministry.

> And Jesus entered the temple and drove out all who sold and bought in the temple, and he overturned the tables of the money-changers and the seats of those who sold pigeons. He said to them, "It is written, 'My house shall be called a house of prayer,' but you make it a den of robbers." And the blind and the lame came to him in the temple, and he healed them. (Matt 21:12–14)

It is clear that Jesus understood the purpose of the temple to include all people regardless of their biology. There are three specific kinds of biology which would have excluded people from the temple and which Jesus refutes in this passage: ethnicity, sexual biology, and physical disability. In saying, "It is written, 'My house shall be called a house of prayer,'" Jesus referred to Isaiah 56:7c: "My house shall be called a house of prayer for all peoples." The wider context of Isaiah 56 explains to whom this applies—the eunuch and the foreigner.

> For thus says the LORD: "To the eunuchs who keep my Sabbaths, who choose the things that please me and hold fast my covenant, I will give in my house and within my walls a monument and a

name better than sons and daughters; I will give them an everlasting name that shall not be cut off. And the foreigners who join themselves to the LORD, to minister to him, to love the name of the LORD, and to be his servants, everyone who keeps the Sabbath and does not profane it, and holds fast my covenant—these I will bring to my holy mountain, and make them joyful in my house of prayer; their burnt offerings and their sacrifices will be accepted on my altar; for my house shall be called a house of prayer for all peoples." (Isa 56:4–7)

There are Old Testament restrictions on worship for both eunuchs and foreigners. The eunuch could not enter the temple according to Deuteronomy 23:1—"No one whose testicles are crushed or whose male organ is cut off shall enter the assembly of the LORD." On the basis of genital mutilation and infertility, the eunuch may be the closest biblical category to intersex, although it should not be equated with it. The same chapter in Deuteronomy also bans foreigners from entering the assembly of Israel: "No Ammonite or Moabite may enter the assembly of the LORD. Even to the tenth generation, none of them may enter the assembly of the LORD forever" (Deut 23:3). These restrictions in Deuteronomy 23 are based on biology. But the promise in Isaiah 56 is not based on biology at all. Instead, those who keep the Sabbath, do not profane the covenant, and love the name of the LORD, will be given access to the temple and their worship will be accepted.

Matthew 21:14 also mentions another biological category of people who had been excluded from the temple—the blind and the lame. The blind and the lame were not expected to make the journey to the temple and some Jewish traditions excluded them from the temple altogether (2 Sam 5:8 in the Septuagint—a second-century Greek translation of the Hebrew Bible). Possibly the restrictions on the blind and the lame entering the temple were an extension of the regulations for priests found in Leviticus 21:17–18.[14] But Jesus welcomed the blind and the lame in the temple and healed them.

Jesus' actions in the temple pointed towards Judaism's true goal, in which all people, without restrictions of biology, are able to participate in the true worship of God (John 4:23–24). Yet it was not the cleansing of the temple that ushered in this new freedom, but rather the death of Jesus. In John's gospel, when Jesus cleansed the temple he said, "Zeal for your

14. Keener, *The Gospel of Matthew*, 502.

house will consume me" (John 2:17, quoting Ps 69:9). Psalm 69 is often connected to the death of Christ in the New Testament (e.g., Matt 27:34; Rom 15:3). Therefore, the quotation is a link between the temple cleansing and the death of Jesus. In cleansing the temple, but far more so in his death, Jesus made the way open for true worship. The physical temple is gone; the temple in which worship now takes place is the crucified and risen Jesus.[15]

Therefore, since the death of Jesus has opened access to worship in the most complete way possible without any reference to biology, there is nothing preventing the intersex person from entering into the worshiping community of the church. Whatever Old Testament restrictions once applied because of biology—race, disability, and genitals—are no longer applicable. Consequently, there is no basis for excluding intersex persons who trust in Jesus. If the church is failing to include intersex persons within its worshipping community, then perhaps she has forgotten that the Lord bought her with his own blood.

Jesus Died for the Bride—Sexuality Fulfilled and Healed

Finally, the cross makes possible the fulfilment of human sexuality and brings healing for our broken sexuality. The death of Jesus enables the creation of the church, the bride of Christ, and therefore brings sexuality close to its ultimate goal. In his death, Jesus bore the consequences of broken sexuality, by losing his intimacy with the Father, and therefore opened the way for full healing of our broken sexuality in the resurrection.

The Epistle to the Ephesians uses the metaphor of Christ as the bridegroom and the church as his bride (Eph 5:22–33). Marriage is a mystery that refers to Christ and the church (5:32). This mystery was hidden in the creation account in Genesis 2, and only brought to light in the gospel. In other words, Ephesians 5:32 sheds light on the meaning of Genesis 2:18–25—the creation of the woman for the man.

> Christ loved the church and gave himself up for her, that he might sanctify her, having cleansed her by the washing of water with the word, so that he might present the church to himself in splendour, without spot or wrinkle or any such thing, that she might be holy and without blemish. (Eph 5:25b–27)

---

15. Beasley-Murray, John, 39–40, 42.

The mystery of marriage is that Christ would take the church as his bride. Thus when Jesus came for his bride and brought her into being through the cross, he fulfilled what is intended by the creation of humanity as male and female. We are sexual beings ultimately because of our destiny to be in union with God through Christ (Rev 19:9). It is his death and resurrection that makes that destiny possible.

But Jesus does not merely fulfill what is intended for human sexuality, his death also heals the brokenness of that sexuality. Sexuality is first of all about relationship. Prior to his crucifixion Jesus experienced continual intimacy of relationship with his Father in heaven. It is this unbroken intimacy between Father and Son—who are distinct persons and yet always one in their being—which is the basis for human sexual distinction and fellow humanity. But on the cross something happened to break that never-before-broken intimacy. "And at the ninth hour Jesus cried with a loud voice, '*Eloi, Eloi, lema sabachthani?*' which means, 'My God, my God, why have you forsaken me?'" (Mark 15:34). At that point, Jesus no longer experienced the intimacy with the Father that was always his.

This break in fellowship between Father and Son has implications for sexuality. At the time when Jesus cried out in his apparent abandonment, the unity-in-distinction and distinction-in-unity of Father and Son was utterly unclear to the Son of God, dying on a Roman cross. The basis for sex and gender, the intimacy of relationship between Father and Son, in oneness and distinction, was thus lacking in the cross. There was no basis for human sexuality at that moment. But paradoxically, it was this act of dying for the bride, the church, which fulfilled all that was ever purposed for sex, gender, and human sexuality. In surrendering to the will of the Father to be consumed by evil and death itself, the Son of God revealed, and made real, the mystery that had been hidden through the ages until that point.

Consequently, the death of Jesus actually healed all that is broken in regard to sex, gender, and human sexuality. The cross took Jesus into the depths of human brokenness, and in those depths he healed all that is broken in the human person by bearing it to the uttermost. Since he could not remain dead, he was raised from the dead by the Spirit, revealing that he is the Son of God in the fullest possible sense (Rom 1:4). In Christ there is now no human brokenness, since he has borne that brokenness and emerged as the glorified Son. Now the relationship between the Son and the Father contains not merely the intimacy of God and God, but it contains a human-God relationship, since Jesus is exalted to the right hand of the Father as the God-man. Thus human sexuality was healed in the death and resurrection of Christ. We must await the resurrection in order to see the complete transformation of human sexuality, including physical healing of intersex. But healing of intersex has been won by the death of Christ.

Conclusion

The death of Jesus has immense implications for intersex people. In undergoing persecution and betrayal, he has experienced what many intersex people have gone through in their lives and relationships. In his death Jesus bore the crippling burden of shame that is so frequently attached to being intersex. His death has provided forgiveness for intersex people and he calls hurting intersex persons to forgive those who have hurt them. By losing his identity on the cross Jesus provided a new identity for intersex people as sons of God. Because of the cross, there are no longer any restrictions on worship based on biology. And finally, the death of Jesus brings about the fulfillment of human sexuality and paves the way for the healing of intersex in the resurrection.

But the death of Jesus is by no means the end. He has risen from the dead and his people will also be raised from the dead to join him in glory. Therefore, a theological exploration of intersex cannot end with the cross. In the coming chapter I will consider the future of intersex in the resurrection and the implications of the resurrection for the present.

6

Resurrection and Intersex

Introduction

THE RESURRECTION OF THE dead is the final stage in the human story, the completion of salvation. We have a glimpse of what the resurrection state will be like, because Jesus, the firstfruits of the resurrected dead, has gone before us. Our resurrected bodies will be both like and unlike the bodies we have now. We will genuinely be who we are. But, on the other hand, we will be radically transformed by the resurrection. The resurrected body will be glorious, incorruptible, immortal, and filled to overflowing with the Holy Spirit. The wonder of what we will become through Christ is presently unimaginable.

Human sexuality will also undergo transformation in the resurrection. Christians have maintained through history that humans will continue to be sexual beings when resurrected. Women will be women and men men. But sexuality will not be the same as it is now. Much of what awaits us is as yet unrevealed. Intersex bodies will be healed; intersex people will be restored according to God's creative intent. This is not to say that identity will be in question, since identity is secured in Christ. However, which intersex person will be male and which female cannot be known in the present.

The resurrection of the dead has real implications for the present. The resurrection speaks loudly of the goodness of the body. Since the body is already good, the focus of medical treatment must be on well-being, not cosmetic alteration. A number of aspects of life as an intersex person are difficult and some cannot be changed. The resurrection provides comfort and strength for coping with these things. Lastly, sexual ethics is a product of our hope of resurrection. Intersex-specific sexual ethics, related to marriage and gender, are considered in the final part of the chapter.

The Resurrection of the Dead

The resurrection of the dead will undo death and enable believers to experience God in glorified bodies. When people are raised from the dead they do not become different people, but retain their identities. Yet, the resurrection is something that will completely and radically transform each person. We will become something very different to what we are now, even while remaining the same.

Resurrection, Glory, and Bodies

Although the pressing question about intersex bodies is whether they will be intersex when resurrected, sexuality is actually only one aspect of the resurrected body. Because intersex persons are not defined by their genitals and gonads, but are whole people with other body parts, with relationships, with gifts, talents, and desires, I will begin with a general discussion of the resurrection before moving on to discuss sexuality in the resurrected state.

The first humans sinned and brought death into the world. The rest of the Bible is the history of redemption as the God of Israel chose people and directed them to relationship with himself. God has made a covenant with his people and his faithfulness to the covenant led to redemption for humanity. The resurrection of the dead is the final step in redemption, bringing to completion the goal of human existence. The New Testament calls death the last enemy, which God will finally put beneath Jesus' feet (1 Cor 15:25–26).

The final book of the Bible provides a picture of human destiny as it was truly intended from the time of creation.

> Then I saw a new heaven and a new earth, for the first heaven and the first earth had passed away, and the sea was no more. And I saw the holy city, new Jerusalem, coming down out of heaven from God, prepared as a bride adorned for her husband. And I heard a loud voice from the throne saying, "Behold, the dwelling place of God is with man. He will dwell with them, and they will be his

people, and God himself will be with them as their God. He will wipe away every tear from their eyes, and death shall be no more, neither shall there be mourning, nor crying, nor pain anymore, for the former things have passed away." (Rev 21:1–4)

Those who are redeemed by Christ will live forever in the presence of God, whose dwelling place will be with humanity. There will be no more death and no more of the pain and sorrow that is part of our present existence. The resurrected state will be so glorious that we cannot presently understand how wonderful it will be.

There is an element of mystery about what the bodies of believers will be like. Yet there are some things that can be clearly stated, and these are important for without these we would miss out on the hope set before us in Christ. Primarily, the Bible plainly states that we will be raised with a body. Human beings were created with bodies and we cannot disconnect ourselves from those bodies. In some sense we are our bodies. The Greek idea of the immortal soul is quite different to the biblical account of the resurrection of the dead. We will not live on forever as disembodied souls, leaving behind the difficulties of bodily existence. Rather, "those who are considered worthy to attain to that age and to the resurrection" (Luke 20:35) will be raised in a *body* (Rom 8:23), albeit a body which will be different to the present body in many ways.

Continuity and Discontinuity

Jesus has already been raised from the dead and is now seated at God's right hand (Rom 8:34). The fact that Jesus has already experienced the resurrection means that it is possible to say something about what the resurrection will be like for us. What will the resurrected body be like? There are two important features of the resurrected body: continuity and discontinuity. There is continuity of identity in that the person who is resurrected from the dead is the same person as the one who died. There is no loss of personal identity. We do not become someone else entirely. But there is also discontinuity; the body that is raised is in many ways very different from the body that dies. This is necessarily the case, since on the most basic level a resurrected body is not simply a resuscitated body. A resurrected body will never experience death, and this fact alone makes the new body considerably different from the old.

The continuity between the present body and the future body is evidenced in the resurrected body of Jesus. The tomb of Jesus is now vacated. The body in which he lived and was crucified and died was the body that God raised from the dead. The earthly body of Jesus has been transformed by the resurrection; it is not only the soul of Jesus that was raised, but his body also. This implies that the mortal bodies that we now inhabit are the ones that God will transform in the resurrection of the dead. "If the Spirit of him who raised Jesus from the dead dwells in you, he who raised Christ Jesus from the dead will also give life to your mortal bodies through his Spirit who dwells in you" (Rom 8:11). The body that dies is the body that will be raised from the dead, that is, there is genuine continuity of identity. The resurrected body has authentic connection to the life that was lived by the person. The whole person—body, soul, and spirit—is raised from the dead and will experience a new glorious existence.[1]

The resurrected person is truly the same person who died, and yet that person will be utterly transformed. It is this discontinuity between the person now and the resurrected person that is given expression in the main passage about the resurrection of the dead, 1 Corinthians 15. The resurrection is not simply bringing the body back to life. It involves a complete transformation of the body and bodily existence. This transformation means that Paul cannot explain the resurrected body without resorting to analogies.

The first of these is the connection and disconnection between a seed and a plant.

> And what you sow is not the body that is to be, but a bare kernel, perhaps of wheat or of some other grain. But God gives it a body as he has chosen, and to each kind of seed its own body. (1 Cor 15:37–38)

Although the seed and the plant are clearly connected, since the plant grows out of the seed, they are also clearly different. A plant is not a larger version of the seed. The plant comes into being by being buried, dying, and being transformed; God gives it a new body, different to the original.

The second analogy Paul uses is about different kinds of bodies and different kinds of glory.

---

1. O'Collins, *Believing in the Resurrection*, 141–43; Wright, *The Resurrection of the Son of God*, 256.

> For not all flesh is the same, but there is one kind for humans, another for animals, another for birds, and another for fish. There are heavenly bodies and earthly bodies, but the glory of the heavenly is of one kind, and the glory of the earthly is of another. There is one glory of the sun, and another glory of the moon, and another glory of the stars; for star differs from star in glory. (1 Cor 15:39–41)

This is obviously an argument aimed at a first-century, prescientific audience. But the point is not obscure. Just because the future body will be different to the present body is not a reason to discount its existence. It is clear from our present-day experience that there are different kinds of bodies and different kinds of glory. What our future bodies will be like is not presently conceivable, but we can be sure that we will have a God-given body with a glory appropriate to it.

There is real discontinuity between our present bodies and those that are to come. The seed analogy is continued in regard to the human body:

> So is it with the resurrection of the dead. What is sown is perishable; what is raised is imperishable. It is sown in dishonor; it is raised in glory. It is sown in weakness; it is raised in power. It is sown a natural body; it is raised a spiritual body. If there is a natural body, there is also a spiritual body. (1 Cor 15:42–44)

As with the seed mentioned in verses 37–38, the body sown is not the same as the body raised. The body sown is perishable, without honor, weak, and natural. The body raised is imperishable, glorious, powerful, and spiritual.

Paul hinges the next portion of his argument on the contrast between Adam, the man of dust, and Jesus Christ, the man from heaven (15:47). This distinction is crucial in grasping the disjunction between the old body, which dies, and the new body, which does not. The present body is earthly and doomed to return to the dust. Corruption, decay, death, and complete dissolution are part and parcel of the bodily existence we now have. In contrast, the resurrected body will be part of the new creation. Life in the new creation will involve none of the negatives of the old bodily existence (15:50b, 52b, 53, 54). The physicality of human existence will not cease, since humans are fundamentally embodied creatures. However, resurrection is not going backwards to our primal origins, but rather forwards to the new, which, apart from the resurrected body of Jesus, is still to come.[2]

Although the person who dies is the person who is raised, there are significant differences between our present bodies and our resurrection

2. Wright, *The Resurrection of the Son of God*, 340, 347, 353.

bodies. Given the enormity of this change, it would be inappropriate to insist that if I do not have precisely the same bodily experience and bodily features then my identity will be compromised. There is no doubt that, just as Jesus Christ is the same person who was crucified, believers will rise with identities intact. Yet Jesus has now transcended his earthly existence and subjection to death, and we too will transcend our earthly existence and bondage to death. Our new bodies will be bodies still, but bodies of a different kind, bodies with a greater glory than anything we can presently imagine. We must not to cling too tightly to the body that is perishable, without honor, weak, and natural. Nor should we deny the future transformation of our identities.

The Resurrection and the Sexed Body

With a foundational understanding of the resurrection we can now discuss sexuality in the resurrected state. In so many respects, intersex bodies will be like those of unambiguously sexed persons in the resurrection. But the question remains: will intersex remain intersex in the resurrection? Other people have already considered this issue so I will begin with their opinions and my response. My own argument then follows. This is in three parts: sexuality without marriage, since there is no marriage in the resurrection; the historic Christian tradition regarding the continuity of sexuality in the resurrection; and an argument for the healing of intersex bodies in the resurrection.

Cornwall and DeFranza on Intersex in the Resurrection

Megan DeFranza and Susannah Cornwall have written the two most significant theological books about intersex. Therefore, I will outline here what they have to say about the resurrection of intersex bodies. DeFranza is an evangelical. Her argument is that, in Christ, it is irrelevant whether a person is male, female, or intersex. Revelation 7:9 describes a scene in heaven, in which there is a huge crowd worshipping "from every nation, from all tribes and peoples and languages." In the eschatological community there will be all kinds of people, including eunuchs and intersex persons. These divisions and distinctions are not done away with, because in the eschatological body of Christ difference will no longer matter.[3]

3. DeFranza, *Sex Difference in Christian Theology*, 183–84.

Cornwall is a liberal who questions the perfection of intersex bodies in the resurrection of the dead. Insisting that intersex bodies will be perfected in the end would imply that they are in some way imperfect in the present. This assumption also leaves heteronormativity unquestioned. If we assume that bodies will be healed in the resurrection, this can allow injustices in the present to be ignored. Instead of assuming the future healing of intersex bodies, we must instead seek to question our own issues about bodies in the present. When we think of intersex bodies being perfected in the resurrection, we may conclude that those who are presently unambiguously sexed are somehow closer to salvation than those who are intersex.[4]

According to Cornwall, we cannot conclude too much about impairment from Paul's use of the word "imperishable," because he does not describe the particulars of imperishable bodies. Therefore, uncertainty exists about the extent of continuity between the present and the future body. Instead of perfection, the resurrected body will have a redeemed body story. Augustine argued that the resurrection body will lack nothing from the original body, but have deformities removed. But Cornwall asks how a person with mosaicism can still be the same person if the mosaic genes are removed. She proposes a rereading of Augustine to say that our resurrection bodies will be freed from the meanings that are now ascribed to them.[5]

Cornwall insists that, in light of this future, we must work now to end "erotic domination" (of men over women or of the unambiguously sexed over the intersex), by undoing the structures that enable it. Redemption must begin in the present. Therefore, intersex people must have control over their own bodies. We must perceive these bodies as showing the glory of God in the present. Since Jesus was raised with a wounded body, then it is possible that ambiguous bodies may also be raised ambiguous. Just as we encounter God in the limited body of Jesus, so too we may encounter God is the limited bodies of others.[6]

My Response to DeFranza and Cornwall

These ideas about the resurrected state deserve some response. Both DeFranza and Cornwall have missed the fact that the resurrection affects the whole person, not merely our perceptions of one another. Additionally,

4. Cornwall, *Sex and Uncertainty*, 69, 182–84.

5. Ibid., 185–92.

6. Ibid., 193–96.

Cornwall has an inadequate theology of sin and salvation. Continuity of identity does not exclude transformation in the resurrection. Lastly, she has underplayed the glory of the risen Jesus by projecting the conditions of this world onto the next.

To begin, I borrow some categories from disability studies. There are three generally agreed ways of understanding disability: the medical model, the social model, and the cultural model. The medical model perceives disability only in terms of medical conditions and medical management, assuming that disability resides in the individual body and must be medically rehabilitated. The social model, on the other hand, considers disability as something produced by societal attitudes and restrictions, not something inherent in the person. The third is the cultural model, which understands disability as a product of the way in which cultures narrate and organize the world.[7]

We may fruitfully use these categories in speaking about intersex. A medical model of intersex would see intersex as a medical issue alone. According to this model, the problem lies in the intersex body; the "wayward" body must be corrected. A social model of intersex would consider intersex as something produced, not in bodies, but in social restrictions and discrimination against people with unusual biology. The cultural model might apply in terms of cultural understandings that intersex cannot exist, because humans are only male or female.

Cornwall has applied the social and cultural models to her understanding of the resurrection of the dead. DeFranza has taken a similar stance, emphasizing the transformation of the community, rather than the transformation of intersex bodies. It is without doubt true that when we are raised there will be social and cultural transformation, which will transform the way in which Christians perceive God, the world, and other people. There is no need for us to wait for the resurrection before the church becomes inclusive of all kinds of difference: race, sex, language, and culture. But in ignoring the medical model of intersex, Cornwall and DeFranza have denied that the transformation that takes place in the resurrection will radically alter bodies and individuals as whole people, not just society and culture. In truth, the whole person will be transformed in the resurrection, not merely the way in which we perceive reality.

Cornwall's ideas required a more detailed response. First of all, she cannot accept that intersex bodies are not presently perfect. To her mind,

7. Moss and Schipper, "Introduction," 2–4.

future perfection of bodies implies that those who have bodies closer to the ideal are in some way closer to salvation now. Cornwall has a liberal theological understanding of the world. She has little concept of sin or its consequences. However, the Bible is clear that sin has radically changed the nature of the created world and the humans who live within it. As a consequence, no bodies—whether male, female, or intersex—are perfect now. Every body dies and needs healing in the resurrection. A person's present embodied sexuality is certainly no indication of relationship with God.

Second, Cornwall questions whether a body healed of an intersex condition would preserve the identity of that person. Although there is real continuity of identity in the resurrection, a degree of discontinuity is necessary for the body to become what it will be—immortal, incorruptible, glorious, and spiritual. These things may well be hard to imagine, but are not impossible for him who raised the Lord Jesus from the dead. Our personal identity is secured in our identity as persons "in Christ." No amount of physical transformation can remove that identity.

Third, Cornwall draws on an argument from disability theology, first put forward by Nancy Eiesland, that Jesus was raised in a wounded body.[8] The major reason why Eiesland's conception of the risen Jesus is incorrect is because it does not take into account the new reality into which Jesus has entered. We cannot rightly project the categories and conditions of this present world onto the next. There will be a radical change from the present evil age to the new age. All things will become new; "the former things have passed away" (Rev 21:4–5). The risen Jesus is not constrained by the weaknesses of the earthly life. "He was crucified in weakness, but lives by the power of God" (2 Cor 13:4a).

Cornwall correctly observes that the eschatological future must in some way affect the present. However, her ideas about what that effect should be would not be acceptable to evangelicals, since she has little regard for the sexual ethics outlined in the New Testament. This disagreement aside, the future must impact the present. Later in this chapter I will discuss three important implications of the future resurrection for the present: the goodness of bodies and medical treatment, hope in present difficulties, and sexual ethics.

8. A more extensive critique of Eiesland's understanding of the resurrection can be found in Cox, *Autism, Humanity and Personhood*, Chapter 5.

No Marriage in the Resurrection: What of Sexuality?

Now I offer a careful and biblical alternative to DeFranza and Cornwall. The first part of my discussion addresses the purpose of sexuality in the resurrected state. Human sexuality goes beyond the sexual act and procreation, because men and women are different in many ways. In the resurrected state we will have a radically transformed capacity for communication and hence for communion with others. Perfected in the love of the Holy Spirit, those found "in Christ" will experience the most profound of relationships in the wedding supper of the Lamb. Although this is presently beyond our capacity to grasp fully with our limited perspective, this does not negate the future of human sexuality.

Jesus was asked a tricky question by the Sadducees regarding the resurrection and marriage. Since they did not believe in the resurrection, they thought they could stump Jesus by asking him to sort out the mess of relationships for a woman who had been married seven times, once to each of seven brothers. "In the resurrection, therefore, of the seven, whose wife will she be? For they all had her," asked the Sadducees. But Jesus replied, "You are wrong, because you know neither the Scriptures nor the power of God. For in the resurrection they neither marry nor are given in marriage, but are like the angels in heaven" (Matt 22:28–30).

What, then, is the purpose of human sexuality in the resurrection if there will be no more marriage when people are raised from the dead? Marriage and sexual acts are something given to humanity only for this present age. These are not sinful per se, but they will be superseded in the coming age. This is, however, not the end of human sexuality. Human sexuality consists in more than simply desire for sexual intercourse and the means of procreation. Men and women are different from one another in many ways (just as they are similar in many ways, by virtue of their shared humanity). This difference is evident at birth.[9] No more sexual acts and no more procreation does not imply no more sexual difference. Thus we cannot conclude that the end of marriage will be the end of human sexuality.

We might think about sexuality in the resurrection in terms of communication. Bodies are the sphere in which we communicate and commune with others, but those bodies are restricted by sickness and death.

9. Experimenters have demonstrated a difference in social perception between male and female newborns (Connellan et al., "Sex Differences in Human Neonatal Social Perception.")

The resurrection overcomes these restrictions.[10] On the most fundamental level, sexuality is about relationships, about communication and communion with others. Our ability to communicate and to commune will be maximized in the resurrected body. The resurrection will bring an end to isolation, abuse, misunderstanding, rejection, oppression, and sexism. It is in this capacity to communicate and to commune with others that human sexuality will find its completion.

The distinctions between people will enable enhanced communication even as we experience the fullness of the love of God through the Holy Spirit operating in each one. This is one aspect of Paul's expression "the spiritual body," a body filled with the Holy Spirit. Each person will experience the love of God to its fullest possible extent and share that love with others. There can be no higher or greater expression of human sexuality than this completed relationship of love. Participation in the love of God is the consummation of the God-human relationship, which has been sought by God throughout the history of humanity.

Indeed the resurrection will bring human sexuality to its intended goal. Human sexuality and marriage has a purpose beyond reproduction and companionship, since it points toward the union of God and human beings, or the union of Christ and his church. When the dead are raised, the church will join with Jesus in the wedding supper of the Lamb (Rev 19:9). In this union we will become the sexual beings we were created to be. Existing as sexual beings will then mean something other than people engaging in sexual intercourse and procreation. Instead of the physical intimacy of sexual congress, we will enjoy intimacy with God through Christ, which will far surpass any physical intimacy humans are now capable of. Our bodies will still be sexually differentiated, but our sexual organs will find a different purpose.

From our present vantage point the idea of sexuality continuing in a meaningful way without marriage, sexual intercourse, and reproduction may seem nonsensical. However, this is just a matter of limited perspective. As C. S. Lewis has observed:

> I think our present outlook might be like that of a small boy who, on being told that the sexual act was the highest bodily pleasure should immediately ask whether you ate chocolates at the same time. On receiving the answer "No," he might regard absence of chocolates as the chief characteristic of sexuality. In vain would

10. O'Collins, *Believing in the Resurrection*, 143–44.

you tell him that the reason why lovers in their carnal raptures don't bother about chocolates is that they have something better to think of. The boy knows chocolate: he does not know the positive thing that excludes it. We are in the same position. We know the sexual life; we do not know, except in glimpses, the other thing which, in Heaven, will leave no room for it.[11]

One day the resurrection will no longer be a mystery, but the truth of our own experience. This is something worth waiting for.

The Historic Christian Position on Biological Sex in the Resurrection

The transformation of humanity in the resurrection includes the transformation of human sexuality. Human bodies and human sexuality will not disappear. The historic Christian position on the sexed body in the resurrection is consistent in affirming that male and female will continue on beyond the resurrection of the dead. Men will still be men and women will still be women.

It is interesting to note that the early church fathers argued in opposition to the general philosophical position of the ancient world regarding sex and gender. The ancient world effectively recognized only one sex—male. Women were, according to Aristotle, "misbegotten" males and thus lesser humans. Since virtue and reason were the highest perfections and women were believed to lack these, it was clear to the ancients that women as women were unable to be perfected.[12] The idea that women need to become male in order to experience perfection is also found in the apocryphal *Gospel of Thomas*, which claims that Jesus will make Mary male so that she can enter the kingdom of God.[13]

In contrast to the culture of the day, the early church fathers shared a belief in the resurrection of the body. They also shared a belief that women will rise again *as women* and not as men. Sexual differentiation in resurrected bodies was something that many church fathers insisted on.[14] Early father Augustine observed that being female is not a vice, but something natural to women. In the resurrected state there will be no more child-bear-

11. Lewis, *Miracles*, 260–61.
12. DeFranza, *Sex Difference in Christian Theology*, 116.
13. Loader, *Sexuality in the New Testament*, 115; "The Gospel of Thomas 114".
14. This is discussed at length in Petrey, *Carnal Resurrection*.

ing so the reproductive parts of a woman will find new and beautiful uses. God created both sexes and will thus restore both sexes in the resurrection. It is true that Jesus said that there will be no marriage in the resurrection, but he did not say that there will be no more women. When Jesus said "they neither marry" he referred to men, and "nor are given in marriage" he referred to women. This, Augustine insisted, is evidence that men will be men and women will be women in the resurrection.[15]

The continuity of sexual differentiation in the resurrection is also affirmed by twentieth-century theologian Karl Barth. According to Barth, it is wrong to interpret the statement in Galatians 3:28 that in Christ "there is no male and female" as an indication that male and female will no longer exist after people are raised. Male and female will not come to an end, but only marriage and all the concerns associated with this, such as reproduction and raising children. Being male and female is a very significant part of being human, because this difference enables people to be "fellow humanity." Human sexual differentiation is part of our creaturehood. Therefore, we must expect that in the resurrection male and female sexual distinctions will remain.[16]

Some further arguments can be offered. Humans were created male and female. In the absence of any indication in Scripture that human sexuality will be removed, we must conclude that sexual difference will remain part of human identity in the resurrected state. Human sexuality was declared by God to be "good." It is good in the sense of being good *for* a purpose. Human sexuality can achieve a good purpose of God that cannot be achieved by any other means, even by fellowship with the Creator.[17] In addition, since Jesus remains a gendered human being in his resurrected state, then we must suppose that those who are made to conform to his likeness will similarly remain gendered human beings when they are raised from the dead. Humans as male and female are a picture of the distinction-in-unity and unity-in-distinction of the Father and the Son, held together by the bond of the Holy Spirit. Since this cannot ever come to an end, the human sexual binary will not be undone. For all these reasons, in line with the historic Christian position, we must affirm that human beings will remain sexually differentiated in the resurrection.

---

15. Augustine of Hippo, *The City of God against the Pagans*, XXII.17.

16. Barth, *Church Dogmatics III.2*, 295–96.

17. Heimbach, "The Unchangeable Difference," 281–84.

The Resurrection of Intersex Bodies

Men will be raised as men and women will be raised as women, thus preserving their sexual identity in the resurrection. What, then, are we to make of those people whose sexual identity is presently ambiguous? In other words, will people who are presently intersex remain intersex in the resurrection or will they become male or female? The short answer is that intersex people will become fully male or fully female. This will occur as part of the restoration and completion of human existence and the healing of all human sexuality. These changes will not compromise identity, which is hidden in Christ, and these will not be the only transformation of sexual identity.

In the beginning, "God said, 'Let us make man in our image, after our likeness.' . . . So God created man in his own image, in the image of God he created him; male and female he created them" (Gen 1:26–27). Human sexuality was distorted by the entrance of sin into the world, but the resurrection of the dead implies that the word of God spoken at creation does not return to him empty (Isa 55:11). When we are resurrected, what God intended for his creatures as male and female will finally be brought to fruition. No consequence of human sin can undo the faithful word of the Creator. Therefore, those who presently have intersex bodies will be restored to male or female when raised from the dead.

It may appear as if I am singling out intersex people and assuming that those who are presently unambiguously sexed will not need to be healed. This is not the case. Every human body will be healed in the resurrection, whether unambiguously sexed or intersex. All people have been subject to the ravages of sin and death. Each has been affected in unique physical, emotional, mental, and relational ways and each needs healing. There is every reason to believe that even those people who are presently unambiguously sexed will experience healing of their sexual anatomy. It is difficult, however, to be detailed or specific here, due to the real limitations we face as people on this side still viewing the future "in a mirror dimly" (1 Cor 13:12).

What about continuity of identity? Will intersex persons have identities that no longer represent who they were in the earthly, pre-resurrected life? The answer to this lies in Christ. It is in Christ where we find our fundamental, true identity. Identity is not primarily found in a physical attribute or the shape of our genitals. Every aspect of the human person will be healed in the resurrection of the dead, and that healing must include

the completion of our identity. The intersex body may be the site of confusion due to biology, upbringing, or cultural expectations about being either
male or female.[18] Difficulties with identity may arise simply because other
people want to impose an identity on the intersex individual.[19] However,
the confusion that intersex people now live with will be overcome in the
resurrection.

Even the sexual binary of male and female will undergo a radical
change. Because there will be no more marriage or reproduction, human
sexual organs—both external and internal—will not have the same function as they have in this age. Since many men and women have defined
themselves by their sexual roles and responsibilities in this age, as fathers
and mothers, as husbands and wives, as daughters and sons, these identities will also have to undergo a transformation. Many aspects of our lives,
sexual or otherwise, will be changed by the resurrection. Thus clinging to
the old identity will be impossible. We will at one and the same time be
both the same people who died and also people who are renewed into the
image of Christ. In this way we will be who we are as well as being very
much not like what we once were.

What We Cannot Say and What We Can Say

What will healing look like? What, for example, can be said about the biological sex that will be given to a person with XY chromosomes, born with
micropenis, who was assigned as a girl at birth and had surgery and hormone therapy to maintain this sex? Will this person be a man or a woman
when resurrected? Will a person with mosaicism or a chimera have XX
chromosomes or XY chromosomes when resurrected? And what sex will
the true hermaphrodite have in the resurrection? If I were to attempt to
answer these questions, I believe that my answer would be nothing but un-
founded speculation. The answers to such specific questions must be left in
the category of the yet unknown. "For now we see in a mirror dimly, but
then face to face. Now I know in part; then I shall know fully, even as I have
been fully known" (1 Cor 13:12). This may seem unsatisfactory, but there is
no biblical warrant for saying more.

There is, however, more to this matter than simply the shape of
someone's genitals. The intersex person is not simply genitals or gonads

18. Lev, "Intersexuality in the Family," 42.
19. Stark, "Authenticity and Intersexuality," 271.

or surgical scars; each is a whole person with good aspects and bad, with a life lived either in love or not, with attitudes and likes and dislikes, relationships, work and learning, growth and disappointments. All that is grounded in the person of Jesus will be kept, because this alone is of value. All that was grounded in sin or idolatry outside of Christ will be undone and washed away. The person will be made new. The surgical scars will be healed along with the pain and rejection and shame. The friendships that are in Christ will live on forever. Those things which were done in faith and accomplished in the power of the Holy Spirit will form part of the crown of life (Jas 1:12), which we will in turn lay at the feet of Jesus (Rev 4:10), who is forever blessed. Amen. The whole life of the person who is intersex will become new.

Bodily Goodness, Hope, and Sexual Ethics

The resurrection of the dead will be glorious and it will transform believers in radical ways. But this is in the future. What of the present? What must we do in the interim while we wait for the return of Jesus? The hope that Christians have in the resurrection of the dead has definite implications in the present. Three of these implications will be explored here: the goodness of bodies and medicine, hope in the face of stressful and unsolvable issues, and sexual ethics.

The Goodness of Bodies and Medical Intervention

One of the clearest implications of the resurrection is that bodies are good. The goodness of bodies is affirmed by the resurrection of Jesus and tied to their end state. This will help parents and intersex persons to make medical decisions about the intersex body. Medicine does not need to change intersex bodies to make them acceptable, but should provide appropriate care according to their uniqueness.

The created goodness of human bodies is affirmed above all by the resurrection of Jesus. Jesus Christ is God come in the flesh. He lived an embodied existence and his body died. But God the Father, through the power of the Holy Spirit, raised the body of Jesus from the dead (Rom 8:11; Gal 1:1). Jesus now lives in a resurrected body, which is incorruptible, immortal, and spiritual. In this body Jesus reigns from heaven at the right hand of God and receives worship as the embodied Son of God. Embodiment cannot be other than good if that is the state of the holy Son of God. Thus bodily goodness is continually affirmed throughout eternity by Jesus' embodied existence.

The goodness of human bodies is eschatological, that is, tied to their end state. In the resurrected state there can be no disease or dysfunction in the body; the body is freed from death and all the problems of mortality. Because bodily goodness is derived from this future state, nothing that happens to bodies in the present can alter the goodness of those bodies. No amount of disease, pain, suffering, surgery, illness, or injury can in any way cause a body to be other than good. Cultural expressions of beauty or ugliness cannot render a body "not good." This fact enables believers to be accepting of bodies, both those of others and their own.

The goodness of bodies has implications for medicine in the present. To explore this I will first consider the difference between the practice of medicine as healing and the goals of plastic surgery. Medicine has as its central goal the healing of bodies and promotion of health and well-being.[20] On the other hand, cosmetic surgery seeks to make a change in a perceived bodily defect, improve self-esteem, or achieve a better physical appearance.[21] People undergo cosmetic surgery, not because they are physically sick, but rather in order to look better or younger,[22] because they are unhappy with the way they presently look.[23] Hence cosmetic surgery is not generally done as a means of curing bodily illness. Rather, cosmetic surgery is done because the person does not believe that the body is good as it is.

There is, then, a fundamental difference between medicine used as a means of healing disease, illness, and injury, and medicine used to change something that is not fundamentally harmful to the person. The preservation of the body through medical intervention is a reasonable goal, since bodies are very significant to human identity. But cosmetic surgery is generally not a positive affirmation of the goodness of the body. This distinction may help parents and intersex people to decide on which medical interventions are appropriate as a response to intersex. The intersex body, unusual as it is, is good. It is therefore to be accepted without the need to change its appearance.

20. Miller, Brody, and Chung, "Cosmetic Surgery and the Internal Morality of Medicine," 353.

21. Goodman, "Female Cosmetic Genital Surgery," 155.

22. Huss-Ashmore, "The Real Me," 30.

23. Northrop, *Reflecting on Cosmetic Surgery*, 1.

There are, however, certain conditions and consequences of intersex that need medical treatment. For instance, CAH can be life-threatening and must be medically treated. There are other less life-threatening matters that may still warrant medical intervention. The case of undescended testes in CAIS girls and women is a case in point. The significant potential for cancer developing in the testes may justify the removal of these in order to prevent disease. However, some AIS women want to keep their testes, and taking the risk of cancer is a personal risk, like other choices that people make regarding their own health. Hormone therapy may help with physical well-being. Supposed "sex" hormones actually do a lot more than simply determine secondary sex characteristics. They are important in many aspects of healthy bodily function, affecting diverse bodily organs.[24] Therefore, taking hormones that the body does not naturally make can be important to health and physical well-being.

What about intersex surgery? The resurrection of the dead and the healing of intersex in the resurrection do not imply a need for "corrective" surgery in the present. The intersex body is already good, whatever the appearance of the genitals. It does not need to be altered in order to become acceptable. However, I do not believe that surgery is ruled out completely by this argument. Sometimes surgical intervention can enhance a person's well-being, but that should be left to the personal discretion of the adult intersex person, rather than the decision being made for them as a child.

Like unambiguously sexed people, intersex people should be afforded genuine autonomy over their medical care. According to OII Australia, intersex people have health needs that are sometimes unusual if considered on the basis of their stated gender. Funding for procedures based on whether a medical intervention is warranted for a particular sex can be problematic for intersex people, since their stated sex may not truly describe their biology. For instance, an AIS woman may have a prostate, and it is possible for an intersex man to have ovaries. Currently a great deal of indignity is required to access anti-androgen medication. On the other hand, people who have had gonads removed may need testosterone treatment, which is subject to considerable restriction. OIIA advocates for patient-centered medical treatment, according to actual needs and not according to stated gender.[25] Unusual biology is not a choice or a sin. Medical care must

---

24. Fausto-Sterling, *Sexing the Body*, 147.

25. Admin, "Intersex Health Issues—A Brief Summary."

therefore be provided in a way that upholds the dignity of ambiguously sexed individuals.

Resurrection Hope Experienced in the Present

The goodness of bodies is not the only consequence of the resurrection. Cornwall insists that the work of Christ has brought about justice and healing and that begins now.[26] Although I cannot agree with all that Cornwall suggests in regard to justice, it is true that justice needs to be fought for as a consequence of what we believe about the future hope. Unfortunately, no amount of advocacy is able to undo the past nor is it able to undo some aspects of intersex experience. There are some things that intersex people will have to live with, either because of choices made by others or because the biological facts of intersex are presently unalterable. Since some aspects of being intersex simply cannot be changed in the present, the intersex person needs hope. The genuine hope of future resurrection provides comfort and strength in order to cope with present difficulties.

Bodily appearance can cause emotional trauma. Graham, a man with PAIS, recounts his struggles in regard to his unusual body. Having hypospadias, he could not stand to urinate and was given leave at school to visit the toilet when everyone else had finished. Graham was spared the humiliation of group showers after physical education, so he could hide his different body. But trouble came when at age twelve he began to grow breasts. This struck fear into Graham because, although he would strap his breasts, even this was not enough in situations where he was asked to remove his shirt. This caused him to hide when the likelihood of being shirtless arose. Eventually, at age fourteen, he had a mastectomy.[27]

Some intersex people simply do not want to live with the gender they were assigned at birth. One intersex blogger explains that both he and his wife were born intersex. Both have transitioned to different genders to those assigned at birth. She was assigned male and was never happy with this gender, but was forced to take male hormones as a teenager. Consequently she is very tall and muscular. Her unusual appearance causes frequent public derision and speculation about which sex she is. The blogger began life assigned a female gender and has transitioned to male, but is only five foot two inches tall. The couple is often harassed in public. It is difficult to go

26. Cornwall, "Asking About What Is Better," 382–83.
27. Graham, "Graham."

into public toilets without police being called out by onlookers. Even doctors are unhelpful; the woman has been refused estrogen by her doctor.[28]

People with intersex conditions often experience impaired sexual quality of life. Genitals of intersex people may be unusual in both looks and shape, sometimes making it difficult to engage in heterosexual intercourse. Concerns about what sexual partners may think, or the ability to sexually satisfy a partner, can lower sexual quality of life. Genital surgery, although intended to enable sexual intercourse, may produce subnormal functioning of genitals or may actually reduce the capacity for sexual pleasure. In addition, a large proportion of people with intersex conditions do not have a sexual partner.[29]

Infertility is a common issue for intersex people. Infertility is sometimes a natural consequence of the intersex condition. For instance, AIS women have no womb, and Klinefelter men are often infertile. But infertility is sometimes the result of surgery. Morgan Carpenter, president of OII Australia, claims that involuntary sterilization of intersex persons is a form of torture.[30] Infertility causes the infertile person and spouse profound sadness, anger, and envy of those who are able to have children.[31] It puts intense stress on relationships, on sexual functioning, and on family dynamics. Infertility can sometimes bring a marriage to the breaking point.[32] Such was the case for Tony, a man with PAIS, who observes that his infertility "was the beginning of the end" for his marriage.[33] For others, the inability to have children cements the self-perception of not being a "real" woman.[34]

A very common complaint from those intersex persons who make their feelings public is the unwanted surgery. Morgan Carpenter laments, "Every individual member of OII Australia has experienced some form of non-consensual medical intervention."[35] Once the "normalizing" surgery has been done it cannot be undone. The body of the intersex person is

28. Costello, "Does It Get Better?"
29. Schoenbucher, Schweizer, and Richter-Appelt, "Sexual Quality of Life."
30. Carpenter, "Submission on the Involuntary or Coerced Sterilisation of People with Disabilities in Australia," 16.
31. Wenzel, *Coping with Infertility, Miscarriage, and Neonatal Loss*, 14–17.
32. Peterson et al., "Coping Processes of Couples Experiencing Infertility," 5–6.
33. Tony, "Tony."
34. Alderson, Madill, and Balen, "Fear of Devaluation," 93.
35. Carpenter, "Submission on the Involuntary or Coerced Sterilisation of People with Disabilities in Australia," 1.

permanently changed, even if that person transitions to another gender. Adoptive parents of an intersex child have sued the state of South Carolina for surgically altering the child's genitals. The child's penis was removed at sixteen months old. At eight years old the child identified as a boy.[36] Regardless of the outcome of this lawsuit, there is no undoing this unwarranted and unwanted surgery.

Given that some things simply cannot be changed about the intersex experience, people need to have hope. One of the implications of the resurrection is that there is hope that God will bring about a radical change in the world. When Jesus returns for the church (Col 3:4; 1 Pet 5:4), this present age, with its pain and weakness and grief, will come to an end (Gal 1:4; 1 John 2:17). Then "He will wipe away every tear from their eyes, and death shall be no more, neither shall there be mourning, nor crying, nor pain anymore, for the former things have passed away" (Rev 21:4). The hope of the resurrection provides comfort in the present circumstances.

The comfort of the resurrection in the midst of present suffering is a theme in 2 Corinthians. In the early passages the argument moves from the truth of the resurrection to coping with suffering in the present. The suffering and comfort which Paul speaks about in regard to his own life (1:3–7) are parallel to the life of Jesus, who first suffered and then was comforted in his resurrection from death. The connections that Paul makes between his own suffering and comfort and the death and resurrection of Jesus are even clearer.[37]

> For we do not want you to be unaware, brothers, of the affliction we experienced in Asia. For we were so utterly burdened beyond our strength that we despaired of life itself. Indeed, we felt that *we had received the sentence of death. But that was to make us rely not on ourselves but on God who raises the dead.* (2 Cor 1:8–9, my emphasis)

The late Scottish theologian P. T. Forsyth reminds us that God does more than simply suffer with us in sympathy. He both knows what our problems are and has provided the ultimate solution to those problems. In Christ, God has fully dealt with all evil. Jesus even now is seated in the place of victory, having triumphed over death and evil.

---

36. Southern Poverty Law Center, "Groundbreaking SPLC Lawsuit."

37. Wright, *The Resurrection of the Son of God*, 300.

[O]ur faith is not merely that God is with us, nor that one day He will clear all things up and triumph; but that for Him all things are already triumphant, clear, and sure. All things are working together for good, as good is in the cross of Christ and its saving effect. Our faith is not that one day we shall solve the riddles of providence, and see all things put under us, but that now we see Jesus; and that we commit ourselves to one who has both the solution of every tragic thing and the glory of every dark thing clear and sure in a kingdom that cannot be moved, and, therefore alone, moves for ever on.[38]

The fact that Jesus has triumphed and will cause his people to also triumph can give confidence and comfort to intersex people as they endure the things that cannot be changed.

Sexual Ethics

There are some things that we cannot change, but there are some matters in which we must make a choice. Sexual ethics is a choice that the Christian believer makes to please God. The final part of the chapter is devoted to sexual ethics for the intersex person. Ethical behavior is one consequence of the resurrection (Col 3:1), so it is appropriate that this be addressed here. It is certainly one important question that arises from the experience of intersex. As one intersex woman asks:

Who am I supposed to marry? And why can't I marry the person I love, if that person happens to be a woman? That's crazy. If I really insisted on my intersex-ness . . . if I did kind of wave the intersex flag in the Church, would it be okay for me to marry a man? I look female and I pass as female, I am female. I have XY chromosomes, so on a chromosomal level I am certainly intersex. . . . because I don't have testes, does that make it okay for me to marry a man? If I still had testes though, would it be okay for me to marry a man? Yes? Well, I don't think most people have even begun to think about that.[39]

This question demands an answer. The answer I present here will not satisfy those who want a list of absolute rules for every possible intersex scenario. However, I believe that significant things can be said in keeping with the

38. Forsyth, *The Cruciality of the Cross*, 62.
39. Sarah quoted in "What Do Intersex Christians Say about Intersex and Faith?"

gospel and in keeping with the framework I have built up so far. Here I make suggestions for dealing with a complex issue, which has no direct biblical passages to draw on. However, I am confident that in offering these guidelines I can say with Paul, "I too have the Holy Spirit" (1 Cor 7:40). My suggestions are based on the New Testament teaching that ethics is not a set of rules, but involves a godly life enabled by the Holy Spirit.

With regard to ethics there are two temptations, either moral license or moral strictures. But both positions are the result of living "in the flesh." They both assume that human beings must determine their own ethics instead of accepting that God has enabled ethical living through the power of the Holy Spirit. It is the gift of the Holy Spirit—the guarantee of the new age—which allows Christians to live in anticipation of the life of the resurrection. Instead of being slaves under the tutelage of the law, people in Christ are now sons of God, given the privileges of mature adult sonship. As sons, given freedom by the indwelling Spirit, believers are enabled to make moral responses to life situations (Gal 3:23–4:7).[40]

Although believers are under grace not law (Rom 6:14), moral decisions are not arbitrary. Rather, the Holy Spirit guides and directs Christians to respond correctly to reality. The rule that governs this response is love (Gal 5:6). Love is not an emotion that you feel or a condition in which you desire to have your own needs met by another human being. Love is defined by the actions of Jesus, who "laid down his life for us, and we ought to lay down our lives for the brothers" (1 John 3:16). In other words, love is focused on meeting the needs of the other person. Love is neither dominating nor manipulative, but rather acts only on the basis of valuing the being of the other. With respect to deciding whom to marry, love needs to be exercised in two directions: vertical and horizontal, towards God and towards the other person.

I begin with the vertical. The Christian life is first and foremost a personal relationship with Jesus Christ, who shares his own relationship to the Father with his disciples. Sexual ethics flow out of this relationship. This relationship is centered on the other, so Christians, by virtue of the name, are oriented to pleasing Jesus Christ. The exclusive nature of that relationship means that it is never appropriate to violate the relationship by entering into sexual immorality. To become one flesh with another outside of the marriage bond is to sin against Christ and to grieve the Holy Spirit,

---

40. O'Donovan, *Resurrection and Moral Order*, 22–24.

whose temple we are (1 Cor 6:13–20).[41] Thus Christian sexual ethics are an expression of our love for God.

In a sense there are no particular sexual ethics pertaining to the intersex person. The same sexual ethics apply as those that apply to unambiguously sexed people. As an evangelical I make no apology for expressing a very conservative sexual ethic according to the historical Christian interpretation of the New Testament. Sexual activity is intended to be engaged in only within the confines of marriage. Marriage is intended to be between a man and a woman. Homosexuality, adultery, and sex outside of marriage are outside the will of God. What follows will simply assume this to be the sexual ethic espoused by the writers of the New Testament.

People who are intersex may be unable to engage in penetrative sexual intercourse. This does not, however, reduce them to asexual beings.[42] Therefore, even intersex people whose genitals are malformed or scarred may well engage in sexual activity of different sorts. As sexual beings, actions that are sexual in nature need to be confined to a marriage. If the person wishes to engage in sexual activity, then this should be saved for the spouse of the intersex individual. If marriage is not entered into for whatever reason, the appropriate sexual ethic for the intersex person, as with the unambiguously sexed person in the same situation, is to abstain from all sexual conduct. Celibacy is a valid choice for Christians in preference to fornication. Celibacy does not deny the sexual nature of the person, but expresses devotion to God by putting God's will above sexual desires.

On the horizontal relationship plane the primary issue is one of gender. Since a marriage is biblically construed as between a man and a woman, being either a man or a woman is necessary for entering into marriage. So if an intersex person is to marry in accordance with the biblical view of marriage, then that person must decide on a gender and live out of that gender. For some intersex individuals the decision has been made, because they have accepted the decision made by parents, a decision possibly backed up by surgery in childhood. For others the decision is a serious one. In the most extreme case of a true hermaphrodite, the decision may fall either way.

There is no biblical passage that deals directly with choosing a gender. However, one passage seems readily applicable. In the midst of his advice about marriage in 1 Corinthians 7, Paul tells the Corinthian church:

41. Watson, *Agape, Eros, Gender*, 134–40.
42. Cornwall, "Asking About What Is Better," 379.

> Only let each person lead the life that the Lord has assigned to
> him, and to which God has called him. This is my rule in all the
> churches. Was anyone at the time of his call already circumcised?
> Let him not seek to remove the marks of circumcision. Was
> anyone at the time of his call uncircumcised? Let him not seek
> circumcision. For neither circumcision counts for anything nor
> uncircumcision, but keeping the commandments of God. Each
> one should remain in the condition in which he was called. Were
> you a bondservant when called? Do not be concerned about it.
> (But if you can gain your freedom, avail yourself of the oppor-
> tunity.) For he who was called in the Lord as a bondservant is a
> freedman of the Lord. Likewise he who was free when called is a
> bondservant of Christ. You were bought with a price; do not be-
> come bondservants of men. So, brothers, in whatever condition
> each was called, there let him remain with God. (1 Cor 7:17–24)

Paul applies this advice to marriage in verses 25–27, saying that a person
should remain in the state he or she is presently in, either staying married if
now married or staying unmarried if now unmarried. Yet he insists, "But if
you do marry, you have not sinned, and if a betrothed woman marries, she
has not sinned" (1 Cor 7:28a).

This passage might be applied to the matter of choosing a gender. It is
quite acceptable to live with the gender assigned at birth, and even possibly
scribed into the flesh by surgery. God would not see this as a sin. The situation in which the intersex person finds himself or herself when coming to
know Jesus as Lord is a situation in which that person may validly remain.
However, it is not a sin to transition to the other gender. There is abundant
grace in Christ. A decision about gender for the intersex person should
be made according to grace and with the guidance of the Holy Spirit. The
fact that this is difficult and may take some time to work through is not an
insurmountable problem. Christians should recognize the grace of God in
this matter and support the intersex person in the decision-making process.

Having made the decision about gender, it is then possible to determine whether a romantic relationship can be considered heterosexual or
homosexual. New Testament ethics regarding marriage relationships pre-
suppose a man and a woman. Each has different and non-exchangeable
roles in the marriage.

> Wives, submit to your own husbands, as to the Lord. For the
> husband is the head of the wife even as Christ is the head of the
> church, his body, and is himself its Savior. . . . Husbands, love your

wives, as Christ loved the church and gave himself up for her. (Eph 5:22–23, 25)

Transitioning after marriage would make the marriage unworkable according to these instructions. There cannot be two wives or two husbands. Therefore, once married it would not be appropriate or loving to decide that the gender chosen was wrong and transition to the other gender.

There are some other matters that need attention on the horizontal plane. The intersex person should consider the needs of the potential marriage partner before entering into a serious relationship. A number of topics need to be honestly discussed. Intersex biology may impinge on a successful sexual relationship, and the potential marriage partner should be made aware of these things in advance if they are known. Infertility should also be the subject of discussion before marriage. Many marriages succeed despite the couple being unable to have children. However, love would disclose this matter in advance of the marriage if it is known. The love of a marriage partner may well find these challenges no reason to prevent marriage, but the other person should at least have a choice. Love does not seek its own way (1 Cor 13:5), but puts the needs of the other person before your own needs.

So now I return to the original question: "Who am I supposed to marry?" There are, as discussed above, serious matters to address before entering into any marriage. But being intersex does not mean that the field of marriage partners is open to all possibilities. If an intersex person is to honor both God and a marriage partner, then choices need to be made with the help of prayer and the guidance of the Holy Spirit. These decisions are part of living as people who have been redeemed and given new life in Christ. Love must make choices, choices to love God through ethical decisions in accordance with his word, and choices to love other people through being honest, being consistent, and possibly by renouncing marriage altogether.

Conclusion

The resurrection of the dead is the Christian hope. When the followers of Jesus are raised from the dead they will share in the glory of the risen Savior, who is now exalted and seated at the right hand of God. The resurrection is something so wonderful that we can presently not imagine how glorious it will be. But we can be sure that when this happens the difficulties of this life will be a thing of the past. Intersex bodies will be healed; the pain associated with being intersex in the present will be undone and each intersex believer will be whole in every aspect of their life. The hope of the future resurrection flows into the present lives of believers. Because the resurrection affirms the goodness of bodies, there is no need to alter intersex bodies now. The hope that we have can empower intersex believers to cope with the difficulties they now face. And the grace of God, accompanied by the power of the Holy Spirit, enables ethical choices about sexual behavior.

The resurrection of the dead is the consummation of the salvation that Jesus has wrought for us. There is, however, one more aspect of salvation that has not yet been discussed, that is, the way in which Jesus draws his people together into community. This is the topic of the final chapter.

7

A Call to Include the Other

Introduction

TWO MEN ARE SITTING on the steps of a church. Inside the service is going on; people are singing and praying, having communion, and awaiting the sermon. The two men on the steps get into conversation. One man says to the black man opposite him, "Why aren't you inside with the other people?" The black man responds, "They won't let me in there." Then Jesus replies to the black man, "They won't let me in there either." This tale suggests that the church is often a gathering of exclusion rather than inclusion. "Families who have intersex members are frequently isolated, silenced, and alienated—alone within the Church."[1] They are "our unknown, unwelcomed, unwanted neighbor, whom we have shamed into silence, whom we have left standing at the doorstep, alone."[2] In excluding those for whom Christ died, we exclude Jesus himself.

The church should be a haven, in which intersex Christians can be themselves without hiding and without shame. The purpose of this final chapter is to help Christians to see how our wrong ideas about both God and others prevent us from including intersex people in the family to which they should belong. We have inherited our unloving reactions to other humans from the first sinners. Our false understandings of God do not help the situation and these need to be corrected. The good news is that Jesus

1. Hiebert and Hiebert, "Intersex Persons and the Church," 42.
2. Ibid., 43.

put to death the dividing wall between Jew and Gentile on the cross, and he also overcame the exclusion of otherness within his body. He makes his strength perfect in the weakness of intersex, according to his unfailing mercy and grace. Therefore, the church must embrace the other in order to be complete.

Where We Have Come From: Fear of the Other

Humanity was intended to embrace difference as a gift from God. But Adam and Eve's choice to sin changed the way in which people perceive difference. Instead of Adam and Eve being a community of glorious difference, they became a community of alienation that hated difference. Now, human response to difference often includes domination over others and diminishment of the value of the different. Difference has become something to avoid or quash, both individually and corporately.

Adam and Eve were created different from one another and different from God. The creation was fully of diversity, which Adam and Eve had dominion over. However, the tree of the knowledge of good and evil represented the difference between humans and God and as such was a boundary that was not meant to be crossed. When Eve and then Adam ate of the forbidden fruit they breached that boundary, with extreme consequences. This was the boundary of creaturely position; they desired to become more than creatures, to become equal with God. Breaching that boundary significantly changed the way in which the difference between Adam and Eve was expressed.[3]

When first created, difference was a gift given to Adam through the creation of Eve. Eve reflected Adam back to him, showing him his own limits as a human being. The boundaries of difference delineated the fullness of human being and the need for dependence upon one another. But when they disobeyed, these boundaries were broken. Then difference became perceived as distance from one another. The humans no longer wanted to be bound to God or to one another. Unashamed nakedness gave way to shame. Knowing one another was no longer simple openness, but knowing what should not be known. The first couple was a community, but instead of a community whose differences are gift, they became a community of alienation, even while they had to continue to inhabit the same space.[4]

3. Bantum, "Discipleship and Identity," 143–44.
4. Ibid., 145–46.

The human relationship between what is distant and what is near became distorted and disordered by sin. Instead of humanity respecting the difference between creature and Creator, we want to become like God in ways we have no right to usurp. Difference is no longer valued, but rather presents as a something to be controlled by domination. This domination may be enacted in violence, enslavement, or diminishing the worth of certain bodies. But a more subtle domination finds form in Christian thinking that perceives "difference as deviation from a constructed or false center."[5] Intersex individuals know all too well what it is like to have a body that is unacceptable, devalued, and dominated by others.

This understanding of difference provides some insight into why even Christians fail to embrace people on the margins. As humans, we were created with difference as a gift, but we have come to abhor difference. The false center from which we operate, even within the church, is often our own narcissistic self-consciousness. From the false center "I" define what is good. Whatever or whoever is not like me is therefore bad and must be avoided or forced to conform to my own way of being. This overrides the word of the Creator, which declared all creation good. Repentance is needed to again embrace difference as gift and to renounce the need to dominate the other.

But egocentrism also operates on a social level. Thomas Reynolds calls this the "cult of normalcy." According to Reynolds, the cult of normalcy arises because difference threatens the familiar. Society sets up boundaries to keep out what is different and is thus considered suspicious and frightening. The cult of normalcy polices the normal by insisting that bodies must conform to certain parameters. Stigma is attached to bodies that fail to conform to the ideal, yet almost no one actually lives up to the ideal. The result is that people must lie and pretend in order to be accepted as "normal."[6] Reynolds applies this to marginalizing disability, but his insights apply equally well to intersex. The policing of intersex bodies begins at birth; the lies and pretense go on for a lifetime. The church must see through this policing of bodies for what it is, a denial of difference as gift.

5. Ibid., 146–47.

6. Reynolds, *Vulnerable Communion*, 59–65.

False Ideas about God: Denying the Other

Sin has definitely impacted the way in which humans perceive otherness. But there is a theological issue specific to the Western church that also impacts Christian thinking about the other. Our Western understanding of the Trinity is too shallow and does not consider the three persons of the Trinity as very consequential. As a result we do not regard the needs of the other as important. A better understanding of God is needed.

The problem for Western Christians is that the doctrine of the Trinity seems redundant. Saying anything beyond "God is one" seems unnecessary. It is as if all God's acts in history, including the life of Jesus, have nothing to add to the doctrine of God. The result in the Western church is a modalistic understanding of God.[7] That is to say, Western Christians think of God as one, and then see Father, Son, and Holy Spirit as successive manifestations of the one God, not as three distinct persons. There is a failure to grasp that Father, Son, and Spirit are genuine persons in communion with one another.

This understanding of the Trinity is the theological legacy of Augustine, whose philosophical background made him mistrustful of physical reality. Therefore, God's appearance in physical form in Christ affected Augustine's theology little. Augustine's intellectualizing and spiritualizing of theology distorted his understanding of the nature of God. He accepted the doctrine of the Trinity as true, but explained it in unhelpful ways. For Augustine the essence or substance of God is what matters and the persons are peripheral. The relationships of the persons with one another are not important to their being. Augustine explained the Trinity in terms of the human mind. He thought of the Word as part of the divine mind. The Father is like memory; in him the Trinity is stored. The Spirit is the divine will.[8] The result was a very theoretical understanding of the three divine persons instead of seeing them as genuine persons in relation to one another.

One consequence of Augustine's theology is that the church is conceived of as an institution that ministers grace to individuals, instead of a community, living together in love as the three divine persons live together in love.[9] This concept of Trinity, which ignores the interpersonal relations within God, has produced a church that sees difference as something to be

---

7. Gunton, *The Promise of Trinitarian Theology*, 32.

8. Ibid., 33–48.

9. Ibid., 51.

avoided or marginalized. Since we don't understand that difference begins in God, we consider it unhelpful and undesirable. The church is not so much focused on personal relationships as fulfilling individual desires and needs. As long as the individuals feel that they can fit within the structure, then there is very little concern for the needs of others.

A renewed understanding of the Trinity as persons in communion can change the church so that we treat people differently. Philippians 2:5–11 demonstrates the other-centered nature of God. The Son of God did not hold tight to the privilege of divinity, but took a human nature and became a servant. He allowed himself to be crucified for the salvation of the world. The self-giving did not begin when the Son of God became the human Jesus, it was already present in God. God as Father, Son, and Holy Spirit is a self-giving God. The man Jesus is not grasping and inward focused like the fallen Adam. So those who follow Jesus must similarly be people who look out for the interests of others.

In effect, a more relational understanding of the Trinity, in contrast to the view inherited from Augustine, will help the church to become more community minded instead of individualistic. If difference is not something to be avoided but rather embraced as gift, then each believer will step towards the other instead of away from the other. Community is the way in which God exists. God is not an undifferentiated one, but being in communion. The church is not an undifferentiated and homogenous unit, but many different people who come together to make one body.

The Cross of Christ: Joining Together with the Other

The work of Christ produces a radical change in human nature and thereby enables believers to embrace the other and to perceive difference as gift, rather than seeing the other as an enemy to be avoided or dominated. The indwelling Holy Spirit brings the life, death, and resurrection of Christ to bear upon the Christian, enabling a practical change in thinking, attitude, and action. To demonstrate this point I will explore here the reconciliation between Jew and Gentile, which the death of Jesus made possible. Then I will apply this to unambiguously sexed people and intersex. In Christ we are one.

Reconciliation of human beings with other human beings is one of the fruits of Christ's work. Ephesians 2:11–22 relates reconciliation directly to the work of the cross. Once, the Gentiles were far off from both the nation of Israel and the God of Israel, but now they have been brought near by the blood of Christ (2:12–13). "For [Christ] himself is our peace, who has made us both one and has broken down in his flesh the dividing wall of hostility" (2:14). This reconciliation of Jew and Gentile takes place in Christ. It does not involve either the assimilation of Gentiles into Judaism or Jews becoming Gentiles. Neither Jew nor Gentile can boast about the new society of the church, because all honor goes to Christ.[10] There is no hierarchy of value within the new community; Jew and Gentile are equal.[11] "For through [Christ] we both have access in one Spirit to the Father" (2:18).

Jesus Christ is our peace (Eph 2:14). The biblical idea of peace is primarily about harmonious relationships between persons. Christ is the one who brings about peace between people by taking away enmity in whatever form that appears.[12] His intent was to "create in himself one new man in place of the two, so making peace" (Eph 2:15b). What he created was not the amalgamation of two different peoples, but rather something completely new. This is best translated as "a single new person." It is not that one new *thing* was made out of two old things, but one new *person*. The same idea is found in Galatians 3:28c, "all of you are one in Christ Jesus," where again the Greek implies one person rather than one thing. In dying, the person Jesus Christ created, not a thing, but a person for himself, "the bride of Christ."[13]

The fact that the one new person has been created out of two implies that Christians are permitted to embrace difference and respect their different heritages, enabling all to experience "unity in diversity." At the same time, this stops members of either group demanding special rights or partiality for members of that group. Christians can be tolerant of difference. Even more so, the joining together of the two groups demonstrates that each group needs the other. Salvation is not a matter for individuals, but for people in community.[14] Previously Gentiles had been outsiders to the people of God, but now they are part of the new community. They are also part of God's family (2:19), a place of relational intimacy.[15]

10. Barth, *Ephesians 1–3*, 292.

11. O'Brien, *The Letter to the Ephesians*, 191.

12. Ibid., 193–94.

13. Barth, *Ephesians 1–3*, 308–9.

14. Ibid., 310–11.

15. O'Brien, *The Letter to the Ephesians*, 211.

Although the passage in Ephesians discusses the unity of Jew and Gentile in Christ, it is no less applicable to the unity of other diverse groups within the body of Christ. Others are now joined together into one body and one family. In Christ the "other" becomes a brother or a sister, not someone distant, not a stranger. As members of the household of God and of the body of Christ, it is both possible and fitting for believers to embrace the other in love. Otherness has again become a gift from God. Both the unambiguously sexed and intersex need one another. Intersex believers do not need to pretend to be other than who they are in order to conform to the sexually unambiguous majority. Those who are clearly male or clearly female have no ground for expecting that intersex people change to make the majority comfortable. Instead, both intersex and unambiguously sexed have been reconciled into one body and all can give glory to Jesus Christ, because he has provided this reconciliation. The work of Jesus has rescued us all—male, female, and intersex—from the fear of the other and enabled his people to treat the other as gift.

Strength in Weakness: The Value of the Other

To be intersex is to be in the minority, to be considered unusual at best, or worse still freakish. This is most assuredly a weakness. Weakness seems to be the opposite of good. But the Bible tells us that God's grace is present in our weakness. He uses the weak for his purposes and his glory. Those who are weak experience the power of God and the presence of Christ. As the church, we need to realize our need of the weak, since without them we are incomplete.

Our transformation into the image of Christ requires that we first experience the weakness of the cross before we experience the glory of the resurrection, just as was the case in the life of Jesus (Phil 2:5–11; Rom 8:17). The Apostle Paul understood this. In 2 Corinthians 12:7–8 Paul prayed for his thorn in the flesh to be taken away. But Jesus replied to him, "My grace is sufficient for you, for my power is made perfect in weakness" (2 Cor 12:9a). Paul's response to this revelation is,

> Therefore I will boast all the more gladly of my weaknesses, so that the power of Christ may rest upon me. For the sake of Christ, then, I am content with weaknesses, insults, hardships, persecutions, and calamities. For when I am weak, then I am strong. (2 Cor 12:9b–10)

It would not be a stretch to suggest that having an intersex condition would fall under the heading of a "weakness." Many difficulties in life are experienced by intersex people in relationship to their intersex status. Many of those difficulties are imposed from the outside, but that was the case with Paul's experience. Some of the problems relate to internal feelings of shame and confusion. The source of the difficulties and pain are not as important as the fact that is it precisely in this weakness that the strength of Christ becomes evident.

No one chooses to be intersex. Yet the God and Father of our Lord Jesus Christ is sovereign over all events in life, including intersex conditions. Just as he used women of unusual sexuality to be the ancestors of the Messiah (see chapter 2), so too he uses the weakness of intersex for his glory. It seems contrary to logic that weakness in a believer is a good thing, but the experience of Paul points to just the opposite. When weakness is present so is the power of Christ. Therefore, the weakness of intersex in a believer will be accompanied by the presence of Jesus. As the church, then, we should value the other who is weak, because that weak believer—not weak in faith but weak because of difficulties—carries around the presence of Jesus Christ, who was crucified in weakness but lives by the power of God (2 Cor 13:4).

The value of the weak other is also affirmed in Paul's discussion of the unity in diversity of the body in 1 Corinthians 12.

> The eye cannot say to the hand, "I have no need of you," nor again the head to the feet, "I have no need of you." On the contrary, the parts of the body that seem to be weaker are indispensable, and on those parts of the body that we think less honorable we bestow the greater honor, and our unpresentable parts are treated with greater modesty, which our more presentable parts do not require. (1 Cor 12:21–24a)

Intersex believers may appear to be the least presentable parts of the body. Those who are unambiguously sexed do not consider themselves as the weak ones, but the strong ones, the presentable ones who do not need greater modesty. But the strong and presentable parts of the body need the weak and less presentable parts. In other words, the unambiguously sexed members of Christ's body need the intersex members of Christ's body. God gives more honor to the weak than to the strong. Therefore, to the intersex goes the most honor. We cannot be the body of Christ without intersex believers. Without intersex people the church cannot be complete, whole, or fully what she is intended to be.

For this reason the church must recognize the value of the other. If we fail to see the intersex believer as honored by God in their weakness, then the church will be missing a vital part of the body. Intersex believers are utterly indispensible to the body of Christ. Let us embrace the other in the sure knowledge that we need them and cannot be our true selves without them.

Conclusion

Revisiting the Argument

BEFORE I BEGAN RESEARCHING this book I was largely ignorant of the existence of intersex people. There are many thousands of people who live with this every day and yet they have remained largely hidden to the general populace. In writing this book my desire was to bring the gospel to bear on the issues that surround intersex. The gospel has much to say about intersex. Although the existence of people who are neither clearly male nor clearly female presents a challenge to our understanding of humanity, this has not proved so much of a challenge that it cannot be addressed by the grace and truth which is forever present in Jesus Christ (John 1:14). Before giving a final word to intersex people and to the church, I will sum up the conclusions I have arrived at along the way.

Intersex is a term that describes ambiguous sexual biology. Being intersex is often accompanied by some very negative experiences. One of the major issues has been genital surgery, usually carried out in infancy or early childhood, which serves to cover over the sexual ambiguity so that parents can pretend that it never existed. The result of surgery and secrecy has been a great amount of shame, causing many intersex people to believe that their bodies are not acceptable. Identity is also strained by intersex. All this pain has been going on unnoticed by most people. Even though intersex is becoming better known, many people still confuse it with homosexuality and transgender. But intersex is not a choice, it is not a sin; intersex is simply a biological reality.

Intersex bodies are not something to be ashamed of. The fact that we are created beings makes humans something very precious. Each person, however sexed, is created in the image of God and made because of the love of God. Human bodies are good, fearfully and wonderfully made. The fall of humanity has changed the world, but it has not changed these fundamental aspects of our humanity. In the beginning, humanity was made male and female. But the fall has distorted human sexuality in many ways: emotionally, socially, and physically. Intersex is thus one distortion of human sexuality caused by the fall. However, being unusually sexed does not place a person outside God's sovereign purposes. The unusually sexed were an integral part of God's salvation plan as ancestors of the Messiah.

Intersex is not a valid proof that gender is fluid or that it should be abolished altogether. Gender is of great importance to human beings. It both mirrors the difference-in-unity and unity-in-difference present within God, and also points forward to the marriage of Christ and the church, which is the ultimate destiny of redeemed humanity. By the same token, there is a real place for intersex people in the kingdom of God as followers of Jesus. Jesus Christ is the Savior of all people and calls them to use their gifts in the world and the church. Intersex people can become disciples and grow into the likeness of Christ. The eunuch is even commended by Jesus. Intersex is not a third gender, but intersex people are not diminished in their humanity because of their ambiguous biology.

Jesus demonstrated, through his life and ministry, his genuine concern for intersex people. In becoming human, he identified with intersex people, by his lowly status, his extremely vulnerable childhood, and his experience of being stigmatized. In his ministry, he affirmed the marginalized. Instead of despising those who were unclean, he embraced them. He never judged people on the basis of their biology. Jesus did not reject those who were rejected by society. He gave overwhelming grace to sinners. Jesus would not allow the intersex person to remain marginalized, but would rather have welcomed any who came to him.

In his death, Jesus did more than give honor and dignity to the weak and marginalized; he became marginalized himself. He experienced persecution and betrayal, empathizing with the ambiguously sexed. In his death he bore the shame of intersex in an extreme way. He was abused, dehumanized, and had his genitals exposed to public display. Through forgiveness he calls intersex people to forgive others. His loss of identity provides a new identity to intersex people as sons of God. He has ended all biological restrictions on worship. And lastly, his death for his church makes possible both the ultimate purpose and the healing of human sexuality.

But death is not the end. The resurrection will transform our bodily experience. Instead of the present weakness, perishability, dishonor, and fleshiness, bodies of believers will become glorious, powerful, honored, and spiritual. There will be a corresponding transformation of human sexuality. Although sexuality continues in the resurrected state, much about it will be different. Intersex will be healed. In all this we will still remain who we are, but identity will have its own transformation. That is the future, but the resurrection has implications for the present. It affirms the goodness of the body, helping intersex people make medical decisions. It offers hope to those whose present experience is dogged by difficulties. And it is the foundation for sexual ethics, centered on love and empowered by the Spirit.

Difference was given by God as gift, but humans try to avoid or dominate difference. This is the result of human sin. Christians in the West have an inadequate theology of difference, due to our shallow thinking about the Trinity. But Jesus has united different people into his body through his death. God's presence and power is given to the weak in Christ. Therefore, the church cannot be fully itself until we embrace the other, including intersex people.

A Final Word

I began this book by saying that I was intending to address two particular groups of people: intersex Christians and the church as a whole. Every part of the book has been directed towards both groups of people, but in different ways. Now, at the conclusion of the book, I will address a final word to each group of people separately.

To intersex believers, I want particularly to emphasize that you are acceptable to God without alteration, because you are created in his image, made because of love, and are valued and dignified as an intersex person. No matter that the world is dominated by male and female, there is a real place for you in God's kingdom. God is by no means indifferent to the difficult experiences that you have gone through in life. Although life may be hard, there is hope of true justice in the resurrection. I want to say to you that no matter what has been said and done to you, the God of our Lord Jesus Christ cares about you. Find your identity in Christ, because in Christ is all goodness, dignity, destiny, hope, and purpose.

To the rest of the church I want to say this. Let us remember who we are. The church is the bride of Christ, purchased with his own blood.

The people who make up the bride are those who Jesus redeemed from sin for his own glory. Whatever we are, we are because of grace, not by merit. This is the reason why the church must include within itself those who are on the margins. Let us not be a clique, who only invites in people like ourselves, but instead let us welcome people who are different, who are forgotten, and who are ignored. In so many instances, intersex people are the different, the forgotten, and the ignored. But Jesus calls us to be like him. He did not reject the rejected or cast out the outcast. Jesus embraced the lonely, the ostracized, and the marginalized. If we are the body of Christ in truth, then we will do the same.

Bibliography

"45,X/46,XY Mixed Gonadal Dysgenesis." Orphanet, http://www.orpha.net/consor/cgi-bin/OC_Exp.php?Lng=GB&Expert=1772.

"About OII Australia." OII AustraliaIntersex Australia, https://oii.org.au/information/about/.

Admin. "Intersex Health Issues—A Brief Summary." OII Australia, https://oii.org.au/21019/intersex-health-issues/.

AISSG. "What Is AIS?" Androgen Insensitivity Syndrome Support Group, http://www.aissg.org/21_OVERVIEW.HTM.

AISSGA. "Androgen Insensitivity Syndrome: Support and Information for Those Affected by Androgen Insensitivity Syndrome (AIS) and Similar Conditions." http://aissga.org.au/AIS_brochure.pdf.

———. "Peer Support, Information and Advocacy for Intersex People and Their Families." AISSGA, http://aissga.org.au/aims.htm.

"Alberta Students to Define Their Own Gender: 5 New School Guidelines." CBC News, http://www.cbc.ca/news/canada/calgary/alberta-education-gender-lgbtq-gsa-guidelines-1.3402300.

Alderson, Julie, Anna Madill, and Adam Balen. "Fear of Devaluation: Understanding the Experience of Intersexed Women with Androgen Insensitivity Syndrome." *British Journal of Health Psychology* 9.1 (2004) 81–100.

Anderson, Janice Capel, and Stephen D. Moore. "Matthew and Masculinity." In *New Testament Masculinities*, edited by Stephen D. Moore and Janice Capel Anderson, 67–91. Atlanta, GA: Society of Biblical Literature, 2003.

Augustine of Hippo. *The City of God against the Pagans*.

Bantum, Brian. "Discipleship and Identity: A Theological Consideration of Race, Gender, and the Human Situation." In *Sex, Gender, and Christianity*, edited by Priscilla Pope-Levison and John R. Levison, 140–59. Eugene, OR: Wipf and Stock, 2012.

Barbaro, M., A. Wedell, and A. Nordenström. "Disorders of Sex Development." *Seminars in Fetal and Neonatal Medicine* 16 (2011) 119–27.

Barger, Lillian Calles. *Eve's Revenge: Women and a Spirituality of the Body*. Grand Rapids: Brazos, 2003.

Barth, Karl. *Church Dogmatics III.2 The Doctrine of Creation*. Translated by Harold Knight, G. W. Bromiley, J. K. S. Reid and R. H. Fuller. Edinburgh: T&T Clark, 1960.

———. *Church Dogmatics III.4 The Doctrine of Creation.* Translated by A. T. Mackay, T. H. L. Parker, H. Knight, H. A. Kennedy and J. Marks. Peabody, MA: Hendrickson, 1961.

Barth, Markus. *Ephesians 1–3: Introduction, Translation, and Commentary on Chapters 1–3.* The Anchor Bible. Garden City, NY: Doubleday, 1974.

Beasley-Murray, George R. *John.* Word Biblical Commentary. Waco, TX: Word, 1987.

Bereshit Rabba.

Berkouwer, G. C. *Man: The Image of God.* Grand Rapids: Eerdmans, 1962.

Blackless, Melanie, Anthony Charuvastra, Amanda Derryck, Anne Fausto-Sterling, Karl Lauzanne, and Ellen Lee. "How Sexually Dimorphic Are We? Review and Synthesis." *American Journal of Human Biology* 12.2 (2000) 151–66.

Blinzler, Josef. *The Trial of Jesus.* New York: Newman, 1959.

Bonhoeffer, Dietrich. *Creation and Fall: A Theological Exposition of Genesis 1–3.* Translated by Martin Rüter and Ilse Tüdt. Dietrich Bonhoeffer Works 3. Minneapolis: Fortress, 1997.

Borg, Marcus J., and N. T. Wright. *The Meaning of Jesus: Two Visions.* San Francisco: Harper Collins, 1999.

Briffa, Tony. "About Tony." http://briffa.org/about.

Brunner, Emil. *Man in Revolt: A Christian Anthropology.* Translated by Olive Wyon. Philadelphia: Westminster, 1939.

Burk, Denny. *What Is the Meaning of Sex?* Wheaton, IL: Crossway, 2013.

Callahan, Gerald N. *Between XX and XY: Intersexuality and the Myth of Two Sexes.* Chicago: Chicago Review, 2009.

Calvin, John. *Institutes of the Christian Religion.* Translated by Ford Lewis Battles. London: SCM, 1960 [1559].

Cameron, D. "Caught Between: An Essay on Intersexuality." In *Intersex in the Age of Ethics*, edited by Alice Domurat Dreger, 91–96. Hagerstown, MD: University Publishing Group, 1999.

Carpenter, Morgan. "Submission on the Involuntary or Coerced Sterilisation of People with Disabilities in Australia." 2013.

Chase, Cheryl. "Affronting Reason." In *Looking Queer: Body Image and Gay Identity in Lesbian, Bisexual, Gay, and Transgender Communities*, edited by Dawn Atkins, 205–19. New York: Harrington Park, 1998.

———. "Surgical Progress Is Not the Answer to Intersexuality." In *Intersex in the Age of Ethics*, edited by Alice Domurat Dreger, 147–59. Hagerstown, MD: University Publishing Group, 1999.

———. "What Is the Agenda of the Intersex Patient Advocacy Movement?" In *First World Congress: Hormonal and Genetic Basis of Sexual Differentiation Disorders*, 1–7. Tempe, AZ: 2002.

Connellan, Jennifer, Simon Baron-Cohen, Sally Wheelwright, Anna Batki, and Jag Ahluwalia. "Sex Differences in Human Neonatal Social Perception." *Infant Behavior and Development* 23.1 (2000) 113–18.

Conway, Colleen M. "'Behold the Man': Masculine Christology and the Fourth Gospel." In *New Testament Masculinities*, edited by Stephen D. Moore and Janice Capel Anderson, 163–80. Atlanta, GA: Society of Biblical Literature, 2003.

Cornwall, Susannah. "Asking About What Is Better: Intersex, Disability, and Inaugurated Eschatology." *Journal of Religion, Disability & Health* 17.4 (2013) 369–92.

———. "British Intersex Christians' Accounts of Intersex Identity, Christian Identity and Church Experience." *Practical Theology* 6.2 (2013) 220–36.

———. *Sex and Uncertainty in the Body of Christ: Intersex Conditions and Christian Theology.* London: Equinox, 2010.

———. "Telling Stories about Intersex and Christianity: Saying Too Much or Not Saying Enough?" *Theology* 117.1 (2014) 24–33.

Costello, Cary Gabriel. "Does It Get Better?" In *The Intersex Roadshow,* http://intersexroadshow.blogspot.com.au/2010/10/comment-on-wear-purple-day.html.

Cousar, Charles B. *Philippians and Philemon: A Commentary.* Louisville: Westminster John Knox, 2013.

Coventry, Martha. "Finding the Words." In *Intersex In the Age of Ethics,* edited by Alice Domurat Dreger, 71–76. Hagerstown, MD: University Publishing Group, 1999.

Cox, Jennifer Anne. *Autism, Humanity and Personhood: A Christ-Centred Theological Anthropology.* Cambridge: Cambridge Scholars, 2017.

Creighton, Sarah, and Catherine Minto. "Managing Intersex." *British Medical Journal* 323.7324 (2001) 1264–65.

Crouch, Andy. "Sex Without Bodies: The Church's Response to the LGBT Movement Must Be That Matter Matters." *Christianity Today* 57.6 (2013) 74–75.

Crouch, Robert A. "Betwixt and Between: The Past and Future of Intersexuality." In *Intersex in the Age of Ethics,* edited by Alice Domurat Dreger, 29–49. Hagerstown, MD: University Publishing Group, 1999.

"Danvers Statement." The Council for Biblical Manhood and Womanhood, http://cbmw.org/about/danvers-statement/.

DeFranza, Megan K. *Sex Difference in Christian Theology: Male, Female, and Intersex in the Image of God.* Grand Rapids: Eerdmans, 2015.

Devore, Howard. "Growing Up in the Surgical Maelstrom." In *Intersex in the Age of Ethics,* edited by Alice Domurat Dreger, 79–81. Hagerstown, MD: University Publishing Group, 1999.

Dewhurst, John, and Ronald Gordon. *The Intersexual Disorders.* London: Baillière Tindall & Cassell, 1969.

Diagnostic and Statistical Manual of Mental Disorders: DSM–5. Arlington, VA: American Psychiatric Association, 2013.

Dowling, Tim. "The Swedish Parents Who Are Keeping Their Baby's Gender a Secret." *The Guardian,* http://www.theguardian.com/lifeandstyle/2010/jun/22/swedish-parents-baby-gender.

Dreger, Alice Domurat. "A History of Intersex: From the Age of Gonads to the Age of Consent." In *Intersex in the Age of Ethics,* edited by Alice Domurat Dreger, 5–22. Hagerstown, MD: University Publishing Group, 1999.

———. "Intersex and Human Rights: The Long View." In *Ethics and Intersex,* edited by Sharon E. Sytsma, 73–86. Dordrecht: Springer, 2006.

Evans, Rachel Held. "The False Gospel of Gender Binaries." http://rachelheldevans.com/blog/gender-binaries.

Farley, Suzanne J. "Challenging the Disorders of Sex Development Dogma." *Nature Reviews Urology* 8.1 (2011) 5.

Fausto-Sterling, Anne. "The Five Sexes, Revisited." *Sciences* 40.4 (2000) 18–23.

———. "The Five Sexes: Why Male and Female Are Not Enough." *The Sciences* 33.2 (1993) 20–24.

————. *Sexing the Body: Gender Politics and the Construction of Gender*. New York: Basic, 2000.

Feder, Ellen K. *Making Sense of Intersex: Changing Ethical Perspectives in Biomedicine*. Bloomington, IN: Indiana University Press, 2014.

Forsyth, P. T. *The Cruciality of the Cross*. Blackwood, SA: New Creation, 1994 [1909].

Franke, John R. "God is Love: The Social Trinity and the Mission of God." In *Trinitarian Theology for the Church: Scripture, Community, Worship*, edited by Daniel J. Treier and David Lauber, 105–19. Downers Grove, IL: IVP Academic, 2009.

Furtado, Paulo Sampaio, Felipe Moraes, Renata Lago, Luciana Oliveira Barros, Maria Betânia Toralles, and Ubirajara Barroso Jn. "Gender Dysphoria Associated with Disorders of Sex Development." *Natural Review of Urology* 9.11 (2012) 620–27.

Gellman, Jerome. "Gender and Sexuality in the Garden of Eden." *Theology and Sexuality* 12.3 (2006) 391–96.

Giles, Kevin. "Barth and Subordinationism." *Scottish Journal of Theology* 64.3 (2011) 327–46.

————. *The Trinity and Subordinationism: The Doctrine of God and the Contemporary Gender Debate*. Downers Grove, IL: InterVarsity, 2002.

Goodman, Michael P. "Female Cosmetic Genital Surgery." *Obstetrics & Gynecology* 113.1 (2009) 154–59.

"The Gospel of Thomas 114." Early Christian Writings, http://www.earlychristianwritings.com/thomas/gospelthomas114.html.

Gough, Brendan, Nicky Weyman, Julie Alderson, Gary Butler, and Mandy Stoner. "'They Did Not Have a Word': The Parental Quest to Locate a 'True Sex' for Their Intersex Children." *Psychology and Health* (2008) 493–507.

Grabbe, Lester L. *An Introduction to First Century Judaism: Jewish Religion and History in the Second Temple Period*. London: T&T Clark, 1996.

Grabham, Emily. "Citizen Bodies, Intersex Citizenship." *Sexualities* 10.1 (2007) 29–48.

Graham. "Graham." AISSGA, http://aissga.org.au/biographies/graham.htm.

Green, Elizabeth. "More Musings on Maleness: The Maleness of Jesus Revisited." *Feminist Theology* 7.20 (1999) 9–27.

Green, Joel B. *The Gospel of Luke*. New International Commentary on the New Testament. Grand Rapids: Eerdmans, 1997.

Grenz, Stanley J. "Is God Sexual?: Human Embodiment and the Christian Conception of God." In *This Is My Name Forever: The Trinity and Gender Language for God*, edited by Alvin F. Kimel Jr., 190–212. Downers Grove, IL: IVP, 2001.

————. "Theological Foundations for Male-Female Relationships." *Journal of the Evangelical Theological Society* 41.4 (1998) 615–30.

Gross, Sally. "Intersexuality and Scripture." *Theology and Sexuality* 1999.11 (1999) 65–74.

Groveman, Sherri A. "The Hanukkah Bush: Ethical Implications in the Clinical Management of Intersex." In *Intersex in the Age of Ethics*, edited by Alice Domurat Dreger, 23–28. Hagerstown, MD: University Publishing Group, 1999.

Gunton, Colin E. *The Promise of Trinitarian Theology*. 2nd ed. London: T&T Clark, 1997.

————. *The Triune Creator: A Historical and Systematic Study*. Grand Rapids: Eerdmans, 1998.

Hare, John. "Hermaphrodites, Eunuchs, and Intersex People: The Witness of Medical Science in Biblical Times and Today." In *Intersex, Theology, and the Bible: Troubling Bodies in Church, Text, and Society*, 79–96. New York: Palgrave Macmillan, 2015.

Harman, Allan M. "6468 עזר." In *Dictionary of Old Testament Theology and Exegesis Volume 3*, edited by Willem A. VanGemeren, 378–79. Grand Rapids: Zondervan, 1997.

Harrison, Nonna Verna. "Orthodox Arguments against the Ordination of Women as Priests." In *Women and the Priesthood*, edited by Thoma Hopko, 165–88. Crestwood, NY: St Vladimir's Seminary Press, 1999.

Hartley, John E. *Leviticus*. Word Biblical Commentary. Dallas, TX: Word.

Heimbach, Daniel R. "The Unchangeable Difference: Eternally Fixed Sexual Identity for an Age of Plastic Sexuality." In *Biblical Foundations for Manhood and Womanhood*, edited by Wayne Grudem, 275–89. Wheaton, IL: Crossway, 2002.

Hengel, Martin. *Crucifixion: In the Ancient World and the Folly of the Message of the Cross*. London: SCM, 1977.

Herndon, April. "Do Frequency Rates Matter?" Intersex Society of North America, http://www.isna.org/node/972.

Hester, David. "Intersexes and the End of Gender: Corporeal Ethics and Postgender Bodies." *Journal of Gender Studies* 13.3 (2004) 215–25.

Hester, David J. "Intersex and the Rhetoric of Healing." In *Ethics and Intersex*, edited by Sharon E. Sytsma, 47–71. Dordrecht: Springer, 2006.

Hiebert, Valerie, and Dennis Hiebert. "Intersex Persons and the Church: Unknown, Unwelcomed, Unwanted Neighbors." *The Journal for the Sociological Integration of Religion and Society* 5.2 (2015) 31–44.

Hodges, Frederick M. "The Ideal Prepuce in Ancient Greece and Rome: Male Genital Aesthetics and Their Relation to *Lipodermos*, Circumcision, Foreskin Restoration, and the *Kynodesmē*." *Bulletin of the History of Medicine* 75.3 (2001) 375–405.

Holmes, Morgan. *Intersex: A Perilous Difference*. Selinsgrove, PA: Susquehanna University Press, 2008.

Houston, James. *I Believe in the Creator*. Edited by Michael Green. London: Hodder and Stoughton, 1979.

Howe, Edmund G. "Intersexuality: What Should Care Providers Do?" In *Intersex in the Age of Ethics*, edited by Alice Domurat Dreger, 211–23. Hagerstown, MD: University Publishing Group, 1999.

Hsu, Al. *The Single Issue*. Leicester: IVP, 1997.

Hughes, Ieuan A. "Disorders of Sex Development: A New Definition and Classification." *Best Practice & Research Clinical Endocrinology & Metabolism* 22.1 (2008) 119–34.

Huss-Ashmore, Rebecca. "The Real Me: Therapeutic Narrative in Cosmetic Surgery." *Expedition: The Magazine of the University of Pennsylvania* 42.3 (2000) 26–37.

Hutson, J. M. "Cliteral Hypertrophy and Other Forms of Ambiguous Genitalia in the Labour Ward." *Australia and New Zealand Journal of Obstetrics and Gynaecology* 32.3 (1992) 238–39.

Iqbal, Farida. "The Persecution of Caster Semenya—Sport and Intersex People's Rights." Links: International Journal of Socialist Renewal, http://links.org.au/node/1266.

Ismail, Ida, and Sarah Creighton. "Surgery for Intersex." *Reviews in Gynaecological Practice* 5 (2005) 57–64.

ISNA. "Androgen Insensitivity Syndrome (AIS)." Intersex Society of North America, http://www.isna.org/faq/conditions/ais.

———. "Frequently Asked Questions." Intersex Society of North America, http://www.isna.org/faq/printable.

———. "Our Mission." http://www.isna.org/.

Jenson, Robert W. *The Triune Identity: God According to the Gospel.* Eugene, OR: Wipf and Stock, 1982.

Jewett, Paul K. *Man as Male and Female: A Study in Sexual Relationships from a Theological Point of View.* Grand Rapids: Eerdmans, 1975.

Jones, Peter R. "Sexual Perversion: The Necessary Fruit of Neo-Pagan Spirituality in the Culture at Large." In *Biblical Foundations for Manhood and Womanhood,* edited by Wayne Grudem, 257–73. Wheaton, IL: Crossway, 2002.

Jones, Stanton L., and Mark A. Yarhouse. "Anthropology, Sexuality, and Sexual Ethics: The Challenge of Psychology." In *Personal Identity in Theological Perspective,* edited by Richard Lints, Michael S. Horton, and Mark R. Talbot, 118–36. Grand Rapids: Eerdmans, 2006.

Jung, Patricia Beattie. "Christianity and Human Sexual Polymorphism: Are They Compatible?" In *Ethics and Intersex,* edited by Sharon E. Sytsma, 293–309. Dordrecht: Springer, 2006.

Juul, Anders, Katharina M. Main, and Neils E. Shakkebaek. "Disorders of Sex Development—The Tip of the Iceberg?" *Natural Review of Endocrinology* 7.9 (2011) 504–5.

Karkazis, Katrina. *Fixing Sex: Intersex, Medical Authority, and Lived Experience.* Durham, NC: Duke University Press, 2008.

Keener, Craig S. *The Gospel of Matthew: A Socio-Rhetorical Commentary.* Grand Rapids: Eerdmans, 2009.

Kemp, Stephen F. "The Role of Genes and Hormones in Sexual Differentiation." In *Ethics and Intersex,* edited by Sharon E. Sytsma, 1–16. Dordrecht: Springer, 2006.

Kessel, Edward L. "A Proposed Biological Interpretation of the Virgin Birth." *Journal of the American Scientific Affiliation* 35 (1983) 129–36.

Kessler, Suzanne J. *Lessons from the Intersexed.* New Brunswick, NJ: Rutgers University Press, 2002.

Kim, Grace Ji-Sun. "Revisioning Christ." *Feminist Theology* 10.28 (2001) 82–92.

Kim, Kun Suk, and Jongwon Kim. "Disorders of Sex Development." *Korean Journal of Urology* 53.1 (2012) 1–8.

Koyama, Emi. "What is Wrong with 'Male, Female, Intersex.'" Intersex Initiative, http://www.intersexinitiative.org/articles/letter-outsidein.html.

Lebacqz, Karen. "Difference or Defect? Intersexuality and the Politics of Difference." *The Annual of the Society of Christian Ethics* (1997) 213–29.

Lev, Arlene Istar. "Intersexuality in the Family: An Unacknowledged Trauma." *Journal of Gay and Lesbian Psychotherapy* 10.2 (2006) 27–56.

Lewis, C. S. *Miracles.* New York: Harper Collins, 1996.

Loader, William. *Making Sense of Sex: Attitudes towards Sexuality in Early Jewish and Christian Literature.* Grand Rapids: Eerdmans, 2013.

———. *Sexuality in the New Testament: Understanding the Key Texts.* Louisville: Westminster John Knox, 2010.

Lüdemann, Gerd. *Virgin Birth?: The Real Story of Mary and Her Son Jesus.* Translated by John Bowden. Harrisburg, PA: Trinity, 1997.

Luther, Martin. *Luther's Works Volume 1: Lectures on Genesis Chapters 1–5.* Edited by Jaroslav Pelikan. St. Louis: Concordia, 1958.

MacArthur, John F. *Philippians.* MacArthur New Testament Commentary. Chicago: Moody, 2001.

Mathews, Kenneth A. *Genesis 1–11:26*. The New American Commentary. Nashville: Broadman and Holman, 2002.

McCredie, Jane. *Making Girls and Boys: Inside the Science of Sex*. Sydney, NSW: University of New South Wales Press, 2011.

Melcher, Sarah J. "A Tale of Two Eunuchs: Isaiah 56:1–8 and Acts 8:26–40." In *Disability Studies and Biblical Literature*, edited by Candida R. Moss and Jeremy Schipper, 115–28. New York: Palgrave MacMillan, 2011.

"Men, Women and Biblical Equality." Minneapolis: Christians for Biblical Equality, 1989.

Miller, Franklin G., Howard Brody, and Kevin C. Chung. "Cosmetic Surgery and the Internal Morality of Medicine." *Cambridge Quarterly of Healthcare Ethics* 9.3 (2000) 353–64.

Mollenkott, Virginia Ramey. *Omnigender: A Trans-Religious Approach*. Cleveland, OH: Pilgrim, 2007.

Morena, Angela. "In Amerika They Call Us Hermaphrodites." In *Intersex in the Age of Ethics*, edited by Alice Domurat Dreger, 137–39. Hagerstown, MD: University Publishing Group, 1999.

Morris, Leon. *Luke: An Introduction and Commentary*. Tyndale Commentary on the New Testament. Grand Rapids: Eerdmans, 1988.

Moss, Candida R., and Jeremy Schipper. "Introduction." In *Disability Studies and Biblical Literature*, edited by Candida R. Moss and Jeremy Schipper, 1–12. New York: Palgrave Macmillan, 2011.

Muram, David, and John Dewhurst. "Inheritance of Intersex Disorders." *Canadian Medical Association Journal* 130.2 (1984) 121–25.

Northrop, Jane Megan. *Reflecting on Cosmetic Surgery: Body Image, Shame and Narcissism*. London: Routledge, 2012.

NSW Government Health: Centre for Genetic Education. "Fact Sheet 31: Klinefelter Syndrome." 2013.

O'Brien, Peter T. *The Epistle to the Philippians: A Commentary on the Greek Text*. Grand Rapids: Eerdmans, 1991.

———. *The Letter to the Ephesians*. The Pillar New Testament Commentary. Grand Rapids: Eerdmans, 1999.

O'Collins, Gerald. *Believing in the Resurrection: The Meaning and Promise of the Risen Jesus*. New York: Paulist, 2012.

O'Donovan, Oliver. *Resurrection and Moral Order: An Outline for Evangelical Ethics*. Leicester: InterVarsity, 1986.

Ozar, David T. "Towards a More Inclusive Conception of Gender-Diversity for Intersex Advocacy and Ethics." In *Ethics and Intersex*, edited by Sharon E. Sytsma, 17–46. Dordrecht: Springer, 2006.

Pannenberg, Wolfhart. *Systematic Theology Volume 2*. Translated by Geoffrey W. Bromiley. Grand Rapids: Eerdmans, 1994.

Peterson, Brennan D., Christopher R. Newton, Karen H. Rosen, and Robert S. Schulman. "Coping Processes of Couples Experiencing Infertility." *Family Relations* 55.2 (2006) 227–39.

Petrey, Taylor Grant. *Carnal Resurrection: Sexuality and Sexual Difference in Early Christianity*. Cambridge, MA: Harvard Divinity School, 2010.

Pinnock, Clark H. *Flame of Love: A Theology of the Holy Spirit*. Downers Grove, IL: InterVarsity, 1996.

Preves, Sharon E. *Intersex and Identity*. New Brunswick, NJ: Rutgers University Press, 2003.

Reiner, William G. "Prenatal Gender Imprinting and Medical Decision-Making: Genetic Male Neonates with Severely Inadequate Penises." In *Intersex and Ethics*, edited by Sharon E. Sytsma, 153–64. Dordrecht: Springer, 2006.

Reis, Elizabeth. *Bodies in Doubt: An American History of Intersex*. Baltimore: Johns Hopkins University Press, 2009.

Reynolds, Thomas E. *Vulnerable Communion: A Theology of Disability and Hospitality*. Grand Rapids: Brazos, 2008.

Roen, Katrina. "Clinical Intervention and Embodied Subjectivity: Atypically Sexed Children and Their Parents." In *Critical Intersex*, edited by Morgan Holmes, 15–40. Burlington, VT: Ashgate, 2009.

Sax, Leonard. "How Common is Intersex? A Response to Anne Fausto-Sterling." *The Journal of Sex Research* 39.3 (2002) 174–78.

Schoenbucher, Verena, Katinka Schweizer, and Hertha Richter-Appelt. "Sexual Quality of Life of Individuals with Disorders of Sex Development and a 46,XY Karyotype: A Review of International Research." *Journal of Sex & Marital Therapy* 36.3 (2010) 193–215.

Sleath, Emma. "I Am Intersex: Georgie Yovanovic's Story." ABC, http://www.abc.net.au/local/stories/2014/12/01/4135509.htm.

———. "I Am Intersex: Shon Klose's Story." ABC, http://www.abc.net.au/local/stories/2014/12/01/4140196.htm.

Slijper, Froukje M. E., Stenvert L. S. Drop, Jan C. Molenaar, and Sabine M. P. F. de Muink Keizer-Schrama. "Long-Term Psychological Evaluation of Intersex Children." *Archives of Sexual Behavior* 27.2 (1998) 125–44.

Southern Poverty Law Center. "Groundbreaking SPLC Lawsuit Accuses South Carolina, Doctors and Hospitals of Unnecessary Surgery on Infant." https://www.splcenter.org/news/2013/05/14/groundbreaking-splc-lawsuit-accuses-south-carolina-doctors-and-hospitals-unnecessary#.UZMv7aKG3Ro.

Stark, Herman E. "Authenticity and Intersexuality." In *Ethics and Intersex*, edited by Sharon E. Sytsma, 271–90. Dordrecht: Springer, 2006.

Steenberg, M. C. *Irenaeus on Creation: The Cosmic Christ and the Saga of Redemption*. Leiden: Brill, 2008.

Stephenson, Lisa P. "Directed, Ordered and Related: The Male and Female Interpersonal Relation in Karl Barth's *Church Dogmatics*." *Scottish Journal of Theology* 61.4 (2008) 435–49.

Sytsma, Sharon E. "The Ethics of Using Dexamethasone to Prevent Virilization of Female Fetuses." In *Ethics and Intersex*, edited by Sharon E. Sytsma, 241–58. Dordrecht: Springer, 2006.

———. "Intersexuality, Cultural Influences, and Cultural Relativism." In *Ethics and Intersex*, edited by Sharon E. Sytsma, 259–70. Dordrecht: Springer, 2006.

———. "Introduction." In *Ethics and Intersex*, edited by Sharon E. Sytsma, xvii–xxv. Dordrecht: Springer, 2006.

Thurston, Bonnie. *Women in the New Testament: Questions and Commentary*. Companions to the New Testament. New York: Crossroad Herder, 1998.

Tony. "Tony." AISSGA, http://aissga.org.au/biographies/tony.htm.

Torrance, James B. *Worship, Community, and the Triune God of Grace*. The Didsbury Lectures. Carlisle: Paternoster, 1996.

Torrance, Thomas F. *Incarnation: The Person and Life of Christ.* Downers Grove, IL: IVP Academic, 2008.

"Tractate Yebamoth Folio 64a."

Trible, Phyllis. *God and the Rhetoric of Sexuality.* Philadelphia: Fortress, 1978.

United Nations High Commissioner for Refugees. *Working with Lesbian, Gay, Bisexual, Transgender and Intersex Persons in Forced Displacement.* Geneva, 2011.

Warne, Garry, and Vijayalakshmi Bhatia. "Intersex, East and West." In *Ethics and Intersex*, edited by Sharon E. Sytsma, 183–205. Dordrecht: Springer, 2006.

Warne, Gary, Maria Craig, Ann Maguire, and Irene Mitchelhill. "Hormones and Me: Congenital Adrenal Hyperplasia (CAH)." Australasian Paediatric Endocrine Group, 2011.

Watson, Francis. *Agape, Eros, Gender: Towards a Pauline Sexual Ethic.* Cambridge: Cambridge University Press, 2000.

Weerakoon, Patricia. *Teen Sex by the Book.* Sydney South, NSW: Youthworks, 2013.

Wenzel, Amy. *Coping with Infertility, Miscarriage, and Neonatal Loss.* Washington, DC: American Psychological Association, 2014.

"What Do Intersex Christians Say about Intersex and Faith?" Lincoln Theological Institute, http://lincolntheologicalinstitute.com/iid-resources.

"Why Not 'Disorders of Sex Development'?" The UK Intersex Association, http://www.ukia.co.uk/ukia/dsd.html.

Wiesmann, Claudia, Susanne Ude-Koeller, Gernot H. G. Sinnecker, and Ute Thyen. "Ethical Principles and Recommendations for the Medical Management of Differences of Sex Development (DSD)/Intersex in Children and Adolescents." *European Journal of Pediatrics* 169 (2010) 671–79.

Wilson, Bruce E., and Williams G. Reiner. "Management of Intersex: A Shifting Paradigm." In *Intersex in the Age of Ethics*, edited by Alice Domurat Dreger, 119–35. Hagerstown, MD: University Publishing Group, 1999.

Wilson, Gina. "Intersex and Religion." In *Heaven Bent: Australian Lesbian, Gay, Bisexual, Transgender and Intersex Experiences of Faith, Religion and Spirituality*, edited by Luke Gahan and Tiffany Jones, 42–46. Melbourne, VIC: Clouds of Magellan, 2013.

———. "Intersex People and Marriage: If We Are Neither Female Nor Male What Exactly Is the Nature of Our Relationship?" Gender Centre, http://www.gendercentre.org.au/resources/polare-archive/archived-articles/intersex-people-and-marriage.htm.

Woodward, Mark, and Nitin Patwardhan. "Disorders of Sex Development." *Surgery (Oxford)* 28.8 (2010) 396–401.

Wright, N. T. *The Resurrection of the Son of God.* London: SPCK, 2003.

Yarhouse, Mark A. *Understanding Gender Dysphoria: Navigating Transgender Issues in a Changing Culture.* Downers Grove, IL: IVP Academic, 2015.

Zucker, Kenneth J. "Gender Identity and Intersexuality." In *Ethics and Intersex*, edited by Sharon E. Sytsma, 165–82. Dordrecht: Springer, 2006.

www.ingramcontent.com/pod-product-compliance
Lightning Source LLC
Chambersburg PA
CBHW030846270326
41928CB00007B/1240